Janet Weil
Oct 2[?]

Philosophical Problems of Quantum Physics

*The NRDC
Natural Resources
Defense Council*

*EDF
Environmental
Defense Fund*

PHILOSOPHICAL PROBLEMS

OF QUANTUM PHYSICS

(Formerly Titled: Philosophic Problems of Nuclear Science)

WERNER HEISENBERG

OX BOW PRESS
Woodbridge, Connecticut

1979 Reprint published by
OX BOW PRESS
P.O. BOX 4045
WOODBRIDGE, CONNECTICUT 06525

Translated by F. C. Hayes
Published in German as *Wandlungen
in den Grundlagen der Naturwissenschaft*

ISBN 0-918024-14-5 (Hardcover)
ISBN 0-918024-15-3 (Paperback)

Library of Congress Card Number 79-89842

Copyright 1952 by Pantheon Books, Inc.

This book was originally published by Pantheon Books, a Divison of Random House, Inc. under the title PHILOSOPHIC PROBLEMS OF NUCLEAR SCIENCE by Werner Heisenberg. This edition is reprinted by arrangement with Pantheon Books, Inc.

Printed in the United States of America

Contents

1. Recent Changes in the Foundation of Exact Science *page* 11
2. On the History of the Physical Interpretation of Nature 27
3. Questions of Principle in Modern Physics 41
4. Ideas of the Natural Philosophy of Ancient Times in Modern Physics 53
5. The Teachings of Goethe and Newton on Colour in the Light of Modern Physics 60
6. On the Unity of the Scientific Outlook on Nature 77
7. Fundamental Problems of Present-day Atomic Physics 95
8. Science as a Means of International Understanding 109
   Index 121

Translator's Note

In preparing this translation the tracing of some of the quotations used by Prof. Heisenberg has been attempted and, in the case of non-German originals, the finding of an accepted translation into English. Passages without source references have been translated directly from the German text.

The term 'atomic physics' has been retained in many places, even when 'nuclear physics' conformed more to current usage in English. It was felt that this would better preserve the link between Greek and modern science as intended by the author.

I wish to express my gratitude to Miss M. English, M.Sc., who has read the typescript, and to several others who have helped with suggestions.

F. C. HAYES.

I
Recent Changes in the Foundations of Exact Science[1]

The development of modern physics, which began with Planck's discovery of the quantum of action and whose intellectual content is expressed in the relativity and quantum theories, has been rounded off in recent years.

The application of the newly discovered principles to further fields of human experience will only become possible when these fields have been subjected to more thorough experimental investigation. On the other hand, an attempt can already be made to sketch a picture of this development, free from the distortions of the controversies of the day, and to clarify the meaning of this development as objectively as possible.

Classical physics, which reached its conclusion some thirty years ago, was built on some fundamental suppositions which appeared to be obvious starting points of all exact science and seemed to require neither proof nor discussion: physics dealt with the behaviour of matter in space and its change in time.

Although this originally characterized only the experiences which formed the basis of physics, nevertheless some properties of matter were inferred from these experiences and seemed to be determined at the same time. One was led to the tacit assumption that there existed an objective course of events in space and time, independent of observation; further, that space and time were categories of classification of all events, completely independent of each other, and thus represented an objective reality, which was the same to all men.

[1] Delivered at the first general session on the occasion of the General Meeting of the 'Gesellschaft deutscher Naturforscher und Aerzte', Hanover, on 17th September, 1934. Originally published in *Naturwissenschaften* 1934, 22 Jahrg., Heft 40.

CHANGES IN THE FOUNDATIONS OF EXACT SCIENCE

The fundamental assumption of classical physics, whose natural consequence was the scientific concept of the universe of the nineteenth century, was first attacked in Einstein's special theory of relativity. I shall here only indicate as much of its fundamentals as is necessary for an understanding of its methodological situation. The rise of this theory was the result of immediate necessity. Classical physics was caught up in contradictions in the attempt to interpret consistently certain subtle experiments, especially the famous experiment of Michelson. Science was forced to admit that one of the assumptions of this classical interpretation was not based on any direct observation since it concerned fields inaccessible to direct observation. This can be compared with the lack of precision inevitable in our daily experience, and the assumption could therefore be dropped. I refer of course to the assumption which delares two events simultaneous even if they do not occur in the same place. We call events 'past' if we can, at least in principle, find out about them through some observation. We call them 'future' if we can still, at least in principle, intervene in their course. It corresponds with our daily experience to believe that events capable of observation are separated from those still open to change only by an infinitely short instant which we call 'present'. This tacit assumption of physics has been proved wrong by the experimental investigations which have forced us to accept the special theory of relativity. In fact there lies between what we have just called 'past' and what we have just called 'future' a small but finite time interval. Its duration is determined by the position of the observer who is deciding on 'past' or 'future' and by the location of the events whose course in time is being investigated. The theory which led to this recognition has meanwhile become an axiomatic foundation of all modern physics, confirmed by a large number of experiments. It has become a permanent property of exact science just as has classical mechanics or the theory of heat. Its extraordinary importance lies in the first place in the completely unexpected realization that a consistent pursuit of classical physics forces a transformation in the very basis of this physics, a situation which we shall frequently meet again. Modern theories did not arise from revolutionary ideas which have been, so to speak, introduced into the exact sciences from without. On the contrary they have forced their way into research which was attempting consistently to carry out the programme of classical physics—they arise out of its very nature. It is for this reason that the beginnings of modern physics cannot be compared with the great upheavals of previous periods like the achievements of Copernicus. Copernicus's idea was much more an import from outside into the concepts of the science of his time, and therefore caused far more telling changes in science than the ideas of modern physics are creating to-day.

The general theory of relativity adds to the revision of the time concept a revision of the geometrical properties of space. If the theory is already correctly interpreting the small number of astronomical observations at present in our possession then, as is well known, a relationship must exist between geometry and the distribution of matter in the universe. Euclid's geometry is then only applicable in small regions of space, while on a large scale space may possess a structure quite different from that immediately obvious to us. The general theory of relativity does not yet rest on such a secure experimental basis as the special theory, though no experiment has definitely contradicted it up to the present. Its convincing power does not lie in the interpretation of many experimental results, which we cannot at present evaluate, but in a new method of thought previously obscured from the view of the scientist. The great new possibilities can be demonstrated clearly by taking the history of the teachings of Copernicus as an example. It is little appreciated nowadays that Copernicus's ideas were, in the beginning, hardly superior to those of Ptolemy in giving a correct presentation of experiences. Again, the proofs which Galileo could muster in support of Copernicus's theories were far less convincing than those at our disposal to-day in support of the general theory of relativity. Nevertheless, the fact that it was not nonsensical to maintain that the earth moved round the sun, proved enough to allow Galileo to mobilize the whole power of his genius behind

Copernicus. In a similar way, the fact that it is not nonsensical to maintain that geometry in the universe depends on the distribution of matter will, independently of any experimental proof, exert such an influence on future research that no theory of gravitation will be able to bypass the general theory of relativity, but instead, will have to absorb it.

Hardly a decade had passed since relativity theory had shown that the basis of exact science, previously looked upon as self-evident, could be changed, when the vital core of classical physics was thrown into doubt by experimental discoveries. There was good reason to disbelieve that the course of an event was objective and independent of the observer. The consequence of these discoveries has led to Bohr's theory of atomic structure. In quantum theory, too, the turning away from the principles of the classical description of nature was *not* effected by the penetration into our science of ideas new and alien to the spirit of earlier physics. On the contrary, science was forced step by step to yield the ground of classical physics as a result of a succession of the most memorable experimental discoveries. After the discovery of the quantum of action by Planck the first and most important step was the recognition (achieved by Lenard's investigations and their interpretation by Einstein) that light, in spite of its wave nature as shown by countless experiments of interference, nevertheless does show corpuscular properties in certain experiments. Thus again we find classical physics, at the beginning of the new theory, involved in inner contradiction when attempting an interpretation of certain experiments entirely consistent with its own principles. In Bohr's atomic theory, which was based on Rutherford's experiments, the dualism alien to classical and earlier physics came even more clearly to the fore. In the succeeding years this theory obtained a firm foundation through a series of experimental and theoretical investigations of which I will mention only those of Franck and Hertz, Stark, Stern and Gerlach on the one hand and those of Sommerfeld, Kramers, Born, Pauli on the other. Subsequently de Broglie discovered also the dualism of wave and corpuscular conceptions in the behaviour of matter. Finally, the simultaneous work of the Göttinger circle and that of Dirac and Schrödinger in fitting the diverse experimental results into a mathematical scheme, established a new and clear situation as regards the principles of physical investigations. An analysis of that situation we owe in the first place to Bohr and this again can only be indicated here. It appears that a peculiar schism in our investigations of atomic processes is inevitable. On the one hand the experimental questions which we ask of nature are always formulated with the help of the plain concepts of classical physics and more especially using the concepts of time and space. For indeed we possess only a form of speech adapted to the objects of our daily environment and capable of describing for instance the structure of some apparatus of measurement. Our experiences, too, can only be made in time and space. On the other hand, the mathematical expressions suitable for the representation of experimental reality are wave functions in multi-dimensional configuration spaces which allow of no easily comprehensible interpretation. Out of this schism there arises the necessity to draw a clear dividing line in the description of atomic processes, between the measuring apparatus of the observer which is described in classical concepts, and the object under observation, whose behaviour is represented by a wave function. Now, while all interrelations on that side of the dividing line leading to the observer, as well as those on the other containing the observed object are sharply defined, (here, by the laws of classical physics and there, by the differential equation of quantum mechanics) the existence of a dividing line is shown in the statistical relationships. The effect of the means of observation on the observed body has to be conceived as a disturbance, partly uncontrolled, in, so to speak, the region of the dividing line. This part of the disturbance, uncontrollable in principle, assumes importance in many different ways. To start with, it is the reason for the appearance of statistical laws of nature in quantum mechanics. Further it imposes a limit on the application of the classical concepts; for the accuracy up to which it is useful to employ these concepts to describe nature intelligibly is limited by the so-called uncertainty relations. It is this very limit of accuracy which indicates the degree of freedom of classical principles essential for any attempt to link up sensibly the various apprehensible explanations—e.g. the concepts of particles and of waves—which fit certain physical phenomena. Finally, this uncontrollable part of the disturbance provides a wonderful method, which can be explored down to the finest detail of fitting together at the dividing line, without contradictions, the fields of the laws of classical and of quantum theory. Thus an entity of laws arises. In this connection it is particularly important that the position of the dividing line—i.e. which objects are to be taken as part of the means of observation and which as part of the observed objects—is immaterial for the purpose of formulating the natural laws. An appreciation of this fact also helps to dispose of an objection frequently made against the finality of quantum mechanics; namely, that behind the interrelations statistically formulated by it, there may be hidden yet another system of determinist natural laws concerning hitherto unknown defining data of nature just as Bolzmann's mechanics of atoms was obscured by the general theory of heat.

A detailed investigation of such a hypothesis shows that these new natural laws would soon be involved in contradictions with the strictly determined results of quantum mechanics. Quantum mechanics allows no room for any addition to its statements, for the only place containing uncertainties is the 'dividing line' previously mentioned. Any attempt to make good the uncertainty of quantum theory, by additions in places determined by certain processes of nature, would entail a shift of the 'dividing line' and would thus bring to light the contradictions between quantum mechanics and the proposed addition.

This immediately raises the more general question of the finality of the changes wrought by modern physics on the foundations of exact science. We have to discuss whether the scientist will once and for all have to renounce all thought of an objective time scale common to all observers, and of objective events in time and space independent of observations on them. Perhaps recent developments represent only a passing crisis. I tend to the opinion, for which there seems to be the strongest evidence, that this renunciation will be final. I would like to begin with an analogy to support this statement. Previous to the beginnings of science in antiquity, the world was conceived as a flat disc, and only the discovery of America and the first circumnavigation of the world destroyed this belief for all time. Of course, nobody had ever seen the edge of the world-disc, but just the same this 'end of the world' acquired form and substance through the legends and imaginings of man. We all know the theme of the ever enquiring man who wants to travel to the end of the world. Then, the question of 'the end of the world' had a definite and clear meaning, but the voyages of discovery of Columbus and Magellan made that question meaningless and transformed the ideas linked to it into fairy tales for ever afterwards. For all that, mankind did not renounce the idea of 'the end of the world' as a result of having explored the whole surface of the world—even to-day there are some unexplored parts—but the voyages of Columbus and Magellan gave clear proofs of the necessity to make use of new lines of approach. In accepting the spherical shape of the earth the loss of the old concept was not felt to be a loss. Similarly modern physics has taught us to do without the concepts of an absolute scale of time and of objective events in space and time. The meaning of these two concepts had never been confirmed by direct experience either, at least not as completely as we had believed. They, too, formed a hypothetical 'end of the world'. It must be stressed that the world of ideas which is to be destroyed simultaneously with these concepts of classical physics is much less living than that destroyed by Columbus or Copernicus. Hence the transition of our concept of the universe, wrought by modern physics, is less decisive than that of the fifteenth and sixteenth centuries. The convincing power of the quantum theory is by no means based on the fact that we may have surveyed all methods of measuring the position and velocity of an electron and that we have been unsuccessful in every case in circumventing the uncertainty relations. But the experimental results of say Compton, Geiger and Bothe are such clear proof of the necessity of making use of the new lines of thought introduced by quantum theory, that the loss of concepts of classical physics no longer appears a loss. The real strength of modern physics, then, rests in its new lines of thought. The hope that new experiments will yet lead us back to objective events in time and space, or to absolute time, are about as well founded as the hope of discovering the end of the world somewhere in the unexplored regions of the Antarctic. This analogy may be further extended; Columbus's discoveries were immaterial to the geography of the Mediterranean countries, and it would be quite wrong to claim that the voyages of discovery of the famous Genoese had made obsolete the positive geographical knowledge of the day. It is equally wrong to speak to-day of a revolution in physics. Modern physics has changed nothing in the great classical disciplines of, for instance, mechanics, optics, and heat. Only the conception of hitherto unexplored regions, formed prematurely from a knowledge of only certain parts of the world, has undergone a decisive transformation. This conception, however, is always decisive for the future course of research.

Proceeding from this short and superficial survey of the most recent developments in theoretical physics we shall now discuss the importance of these events and their possible effects on the future shaping of scientific thought. Science has two tasks: to pass on an understanding of nature, thus enabling man to make nature serve his own purpose, and to indicate to man his appropriate position in nature through a real insight into its inter-relations. The first of these tasks has dominated the development of pure and applied science for the last hundred years and we will therefore discuss it first. The results of theoretical physics, including those of relativity and quantum theory, cannot be made to serve technical progress directly. Theoretical physics exercises its influence on technical development rather indirectly and only after some considerable lapse of time. Two different effects can here be distinguished. First, for an apparatus to be completely adequate for its purpose its design generally presupposes an accurate knowledge of the natural laws involved; e.g. the knowledge of Maxwell's equations, in the form familiar to the engineer and the physicist, is absolutely essential for the construction of a dynamo or a high frequency plant. Similarly a knowledge of the laws of nuclear physics will become essential in the future for the construction of apparatus making use of atomic phenomena. A good deal of time may elapse before these effects of modern physics will have made themselves felt. Secondly, progress in theory may significantly affect the direction of physical research, and thus eventually of technical development. In this connection, the relationship of experimental and theoretical physics must be touched upon. This relation has of late been rather unfairly presented to the German public. It is of course true that experimental work forms, in all fields, the necessary precondition for theoretical understanding and that advances in principle are achieved only under the pressure of experimental results and not by speculation. On the other hand, the direction of experimental work is probably determined by the pointers of theory. The most famous example, since the inception of modern science, of this complementary work determining the relation of theory and experience is the common achievement of Tycho Brahe and Kepler. Tycho's wealth of observational material about planetary motion, which Kepler could never have collected with such accuracy, was the necessary pre-requisite for the latter's work. On the other hand the direction in which astronomy developed in the succeeding centuries was determined by Kepler's discoveries. But we need hardly go back so far to observe the interrelation of experience and theoretical understanding. The transformation of the foundations of exact science that has taken place in modern physics has been brought about step by step as a result of experimental investigations. Yet, a comparison of the fields of investigation in physical laboratories, now and twenty years ago, shows immediately how the direction of experimental research is affected by changes in our understanding of natural laws. Every innovation which exerts its influence on 'observational' physics, in turn effects technical development. Therefore in the present debates as to whether public interest should turn its attention particularly towards engineering, experimental or theoretical science, careful note should be taken that these three branches of science mutually condition and complement one another. The task of pure science at any given time is to clear and prepare the ground for the growth of technical development. Since this ground is quickly taken over, it is important that it should be continually extended and in this theoretical research plays its part. The interaction between technical development and science is in the last resort based on the fact that both spring from the same sources. A neglect of pure science would be a symptom of the exhaustion of the forces which condition both technical progress and science.

However, the effect of the transformations of the foundations of exact science is by no means limited to its influence on technical and experimental research. In the field of the philosophical theory of perception, the beginning of such an effect is being felt. Here the question raised by Kant, and much discussed ever since, has been put into a new light as a result of the critique of absolute time and Euclidean space in relativity theory, and of the laws of causality in quantum theory. I refer to the question of the priority of *Anschauungsformen und Kategorien*. On the one hand it has been shown that our space-time *Anschauungsform* and the laws of causality are not independent of all experience, in the sense that they must remain for the rest of time essential constituents of every physical theory. On the other hand, as Bohr, particularly, has stressed, the applicability of these *Anschauungsformen*, and of the law of causality is the premise of every objective scientific experience even in modern physics. For we can only communicate the course and result of a measurement by describing the necessary manual actions and instrument readings as objective, and as events taking place in the space and time of our *Anschauung*. Neither could we infer the properties of the observed object from the results of measurements if the law of causality did not guarantee an unambiguous connection between the two. The apparent contradictions between these two statements is resolved by means of the following considerations. Physical theories can have a structure differing from classical physics, only when their aims are no longer those of immediate sense perception, i.e. only when they leave the field of common experience dominated by classical physics. It is in this way that modern physics has more accurately defined the limits of the idea of the *a priori* in the exact sciences, than was possible at the time of Kant. There has not yet been a discussion, based on the new outlook, that is sufficiently thorough to show how far this idea is still fruitful in the wider philosophical fields which were essential for Kant.

These special questions of the theory of perception are already connected with the second great problem facing physical theory: that of giving information about the more general interrelations of nature, of which we, ourselves, are part. Science cannot evade this issue if it is to remain true to itself. We need only recollect here, that some of the first representatives of early natural philosophy in antiquity were at the same time centres of religious movements. It is to be expected that the present changes in the scientific concept of the universe will exert their influence upon the wider fields of the world of ideas, when we consider that the changes at the end of the Renaissance transformed the cultural life of the succeeding epochs. The very recent transformations, though not comparable with those at the beginning of modern times, may nevertheless suffice to replace the views, which we may call the scientific concepts of the universe of the nineteenth century, by something new and different. I should like to elaborate this point a little. The scientific views which have become the axiomatic basis of all scientific investigation during the last century, have only assumed gradually, since the beginning of modern times, the rigid forms familiar to us. It was a fundamentally new discovery which gave scientific development its new power. A whole field of reality was found, altogether beyond the appreciation of the Middle Ages during which supernatural revelation was the centre of all thought. Man came up against that objective reality which was free from all doubt and which could be experienced by observation and experiment. The attempt to separate the general from the scientific in objective reality had become the subject of human endeavour, and was a natural consequence of this discovery. A group of axioms emerged from the mass of specific results as the real nucleus of the new science and appeared, as though of a necessity, to be at the root of all scientific investigations. The influence of this new reality was soon exerted in philosophy also, and the foundations of the new understanding of nature appeared as parts of great philosophic systems. Just as, in antiquity, Geometry served as an example of consistency to philosophic thought, so, under the influence of science, were born new philosophic systems. And just as in science, one or several truths recognized as unimpeachable were made the basis of all further deductions, so the same system was used in philosophy, (the systems of Descartes and Spinoza will serve as examples). Even Kant's philosophy, intended as a critique of premature dogmatization in scientific concepts, could not prevent the torpescence of the scientific concept of the universe—it may even be said that it encouraged it. For, once the main reasoning of classical physics had been accepted as the *a priori* of physical investigations, the belief arose, through an obvious though false extrapolation, that it was absolute, i.e. valid for all time, and could never be modified as a result of new experiences.

Thus was formed the solid framework of classical physics, and thus arose the conception of a material world in time and space comparable to a machine which, once set in motion, continues to run, governed by immutable laws. The fact that this machine as well as the whole of science were themselves only products of the human mind appeared irrelevant and of no consequence for an understanding of nature. Only the extension of scientific methods of thought far beyond their legitimate limits of application led to the much deplored division in the world of ideas between the field of science on the one side and the fields of religion and art on the other. Exact science, convinced of the general validity and applicability of scientific principles, interfered in other spheres of intellectual life and thus threatened its own status, Since, however, its power was insufficient to give full expression to these other fields, almost impassable frontiers arose, as in self-defence, between them and science.

The scientific concept of the universe of the nineteenth century, born under these circumstances, is called rational, since its centre, classical physics, can be built up from a small number of axioms capable of rational analysis, and since it rests on belief in the possibility of a rational analysis of all reality. It must, however, be stressed that the hope of gaining an understanding of the whole world from a small part of it can never be supported rationally. Now, the changes in the foundations of science forced upon us by nature in such a marvellous way through atomic phenomena leave classical physics untouched, but they show that scientific systems—like classical mechanics or other parts of classical physics—must always be complete in order to be correct. Hence the extension of scientific investigation to new fields of experience does not mean the application of previously known laws to new subjects. I should like to return again to the analogy I used before, between the discovery of the spherical shape of the earth and the conclusions of modern physics. As long as the earth was taken to be a large disc, there could be hope that the man who had travelled to the end of the world would be able to explain all the things on it. This hope was shattered for ever with the discoveries of Columbus, though they only changed our views about certain parts previously unknown.

Now that we know all our journeying can only bring us back to our starting point, we realise that we are unable to reach full understanding no matter how far we travel. The infinity of the universe lies outside this path. In quite a similar way modern physics has shown that the structure of classical physics—as that of modern physics—is complete in itself. Classical physics extends just as far as the conceptions which form its basis can be applied. But these conceptions already fail us when applied to the processes of nuclear physics, and much more so in the case of all fields of science which are even further removed from classical physics. Thus the hope of understanding all aspects of intellectual life on the principles of classical physics is no more justified than the hope of the traveller who believes he will have obtained the answer to all problems once he has journeyed to the end of the world.

Yet the misunderstanding, that the transformations in exact science have brought to light certain limits to the application of rational thinking, must immediately be countered. A narrower field of application is given to certain ways of thought only, and not to rational thought in general. The discovery that the earth is not the world, but only a small and discrete part of the world, has enabled us to relegate to its proper position the illusory 'end of the world' concept, and instead to map the whole surface of the earth accurately. In a similar way modern physics has purged classical physics of its arbitrary belief in its unlimited application. It has shown that some parts of our science e.g. mechanics, electricity, quantum theory, present scientific systems complete in themselves rational, and capable of complete investigation. They state their respective natural laws, probably correctly, for all time. The essence of this statement is given by the phrase 'completeness in itself' (*Abgeschlossenheit*). The most important new result of nuclear physics was the recognition of the possibility of applying quite different types of natural laws, without contradiction, to one and the same physical event. This is due to the fact that within a system of laws which are based on certain fundamental ideas only certain quite definite ways of asking questions make sense, and thus, that such a system is separated from others which allow different questions to be put. Thus, the transition in science from previously investigated fields of experience to new ones will never consist simply of the application of already known laws to these new fields. On the contrary, a really new field of experience will always lead to the crystallization of a new system of scientific concepts and laws. They will be no less capable of rational analysis than the old ones but their nature will be fundamentally different. It is for this reason that modern physics adopts an attitude very different from classical physics towards all those fields not yet included in its investigations. Let us, for example, consider the problems concerned with the existence of living organisms. From the standpoint of modern physics, according to Bohr, we should expect the laws characteristic of these organisms to be separated from the purely physical laws in a rational and accurately comprehensible manner, just as, say, quantum theory is separated from classical mechanics. A similar solution will, on a smaller scale, apply to the investigations into the properties of the atomic nucleus, which occupies the centre of interest in contemporary physics. The edifice of exact science can hardly be looked upon as a consistent and coherent unit in the naïve way we had hoped. Simply following the prescribed route from any given point will not lead us to all other rooms of this building; for it consists of specific parts, and though each of these is connected to the others by many passageways and each may encompass some others or be encompassed by others, nevertheless each is a unit complete in itself. The advance from the parts already completed to those newly discovered, or to be newly erected, demands each time an intellectual jump, which cannot be achieved through the simple development of already existing knowledge.

Thus contemporary science, to-day much more than at any previous time, has been forced by nature herself to pose again the old question of the possibility of comprehending reality by mental processes, and to answer it in a slightly different way. Previously the example of science could lead to philosophic systems which assumed a certain truth—like the 'cogito, ergo sum' of Descartes—as the starting point from which all questions of 'Weltanschauung' could be attacked. But now nature, through the medium of modern physics has reminded us very clearly that we should never hope for such a firm basis for the comprehension of the whole field of 'things perceptible'. Rather when faced with essentially new intellectual challenges should we continually follow the example of Columbus, who possessed the courage to leave the known world in the almost insane hope of finding land again beyond the sea.

This realization can preserve us from the mistake, not always avoided in the past, of attempting to force new fields of experience into an outmoded, unsuitable structure of concepts.

Conversely, it will therefore be easier to fit methods of thought, originated in contradiction to the ideal of 'complete comprehension' of classical science, into an all-embracing yet unified and logical concept of science. The attempt to link up overhastily the different fields of human knowledge, in the belief that this diversity will perhaps avoid all further difficulties, would effect as little genuine unification of intellectual life as at one time the generalization of rational science led to a rational concept of the universe. But just as that generalization provided for the opening up of new vistas in many fields, so we shall serve the future best by at least easing the way for the newly won methods of thought, rather than by combatting them because of the unfamiliar difficulties they have created. Perhaps it is not too rash to hope that new spiritual forces will again bring us nearer to the unity of a scientific concept of the universe which has been so threatened during the last decades.

2

On the History of the Physical Interpretation of Nature[1]

Exact science of the last thirty years derives its special significance from the fact that its different branches, i.e. Astronomy, Physics and Chemistry have been followed back to their common root—atomic physics. Thus many of the desires which had prompted Leucippus's and Democritus's investigations of nature, have in a certain sense been fulfilled. For that reason it is important, for a deeper understanding of modern science, to find out to what extent present-day research can be treated as a consistent development throughout the centuries of human endeavour for an understanding of nature, and to assess carefully its balance of success and failure. It has become the custom to view the evolution of science as a succession of brilliant and surprising discoveries whose inner connections the human intellect can discover through the instrument of mathematics. Hence it seems important to me to stress, for once, a less apparent tendency which cannot be missed by the careful observer of this evolution, and which is, so to speak, responsible for the inner equilibrium of our science. This is the fact that practically all progress and knowledge in science has been achieved at the sacrifice of previously important formulations of questions and ideas. Thus, an increase in knowledge and perception limits successively the claim of the scientist to an 'understanding' of the world. The observation of nature by man shows here a close analogy to the individual act of perception

[1] This lecture was delivered at the public session of the Academy of Sciences of Saxony on the 19th September, 1932. (Published in the *Ber. d. math.-phys. Klasse*, Bd. 85, 1933.)

which one can, like Fichte, accept as a process of the *Selbstbeschränkung des Ich* (Self-limitation of the ego.) It means that in every act of perception we select one of the infinite number of possibilities and thus we also limit the number of possibilities for the future.

The study of this 'self-limitation', which is part and parcel of all new physical knowledge, gives us some understanding of the degree of compulsion which has determined the path of science in the course of its history. Such a study will at the same time protect modern science against the allegation of partiality and conceit.

The first physical phenomenon to attract the attention of Greek systematic thought, was that of 'substance', of the 'lasting' element in the mutations of all phenomena. In the thesis of Thales, that water is the fundamental substance of which the world 'consists', we can see the formulation of the concept 'matter'. At the very inception of research, none of the words of the preceding clause could, naturally, have a very precise meaning. None of the words 'fundamental substance', 'water' or 'consist' had a concisely defined field of application or an unambiguous meaning, and it was this very fact which gave full freedom to future research. No sacrifice had been made which would limit a unified understanding of the world in the most general sense. Succeeding research defined the term 'fundamental substance' somewhat more concisely. First, it acquired the characteristic of uniformity and indestructibility. This formulation resulted in a complication, for in order to make intelligible the changing phenomena of the world, one had either to assume several fundamental substances whose mixture or separation would account for the innumerable manifestions in our experience, or to separate altogether the concept of the 'lasting' from common experience. Parmenides's idea of the 'being' (*des* '*Seins*') was an attempt in the latter direction. Empedocles regarded earth, fire, air and water as the four 'basic roots' (*Stammwurzeln*) of all things ($ῥιζώματα$). He considered them 'uncreated, indestructible, homogeneous, immutable but at the same time divisible'. Proceeding along the same path, Anaxagoras postulated an infinite number of elements whose joining or separation account for the coming into being and passing away of particular phenomena. This work prepared the ground for an explanation of the qualitative variety of the external world in terms of variations of quantity and changes in the proportions of the mixture and the idea eventually found its consistent conclusion in the atomic theory of Leucippus and Democritus. It recognizes as 'being' only the smallest indivisibile particles of matter, the atoms, whose only quality is the fact that they take up space. The qualitative differences of things perceptible were to be explained by the varying shape, movement and arrangement of atoms in empty space.

This development of the matter-concept from Thales to Democritus undoubtedly represented immense progress in the explanation of the fundamental properties of matter. The possibility of the different states of matter became immediately plausible as did a reasonable interpretation of phenomena connected with the mixing of liquids. Further, as we know to-day, the then unknown concept of chemical compounds received a perspicuous geometrical interpretation. Thus, though we have every reason to admire such progress as a success of consistent development of scientific thought, we must not forget, nevertheless, that these successes necessarily implied a sacrifice of grave importance for the future—I refer to the sacrifice of an 'immediate and direct' understanding of qualities. In our experience qualities like colour, smell and taste are as much immediate and direct realities as shape and movement. In depriving atoms of these qualities—and the very strength of atomic theory lies in this abstraction—one sacrifices immediately the possibility of 'understanding' the qualities of things in the true sense of the word. In place of what we have called 'immediate and direct' understanding' there is in atomic theory a kind of 'analytical' comprehension. The qualities 'red', 'acidic', etc., prove capable of representation in the imagery of certain geometrical and dynamic configurations of atoms. Certain given relations between qualities correspond in experience to obvious geometrical relations of atomic images. The qualitative variety of the world is 'explained' by their reduction to manifold geometric configurations. It can be said, in a sense, by reversing the above statement, that, while Democritus's atomic theory offers an explanation of the qualities mentioned, it still leaves unexplained, i.e. unreduced, the geometrical properties of the world. We must thus distinguish between 'analytical' and 'immediate and direct' concepts. The desire, fulfilled in atomic theory, to depict perceptible qualities of things, like colour and hardness, by means of reduction to geometrical configurations (in the widest sense), enforces the sacrifice of ascertaining the true nature of these qualities by means of science. Thus it can be easily understood why the poets for example always looked upon the atomic concept with horror.

Hand in hand with the development of the concept 'matter' went the attempt to give a more precise meaning to the word 'space' while the naïve conception of the world understood it to consist of many individual things separated by space, the Greek concept of 'empty space' gave rise, at first, to great difficulties in the theory of perception. Parmenides, who had placed the concept of 'being' at the apex of his philosopy, gave it from the very beginning a material character. Existence and taking up space are to him identical. Since there exists only 'being' and as 'non-being' cannot exist, hence empty space (i.e. 'non-being') cannot exist. Parmenides's teaching had, in the last resort, to explain the whole perceptible world as 'imagination'. From it we can feel quite clearly how inconvenient, at first, the concept of empty space must have been to the philosopher. For that reason, a sharp separation of space and its geometrical properties from the concept of matter was not achieved for a considerable time. In Plato's *Timaeus*, for instance, the physical properties of elements are related to geometry, i.e. the properties of space. The individual elements of matter are built up of fundamental components of stereometry and these in turn of simple triangles. Aristotle moved much further than his predecessors, from a deductive science based on abstract principles, to one descriptive and recording. Yet even he brings forward the following proof of the impossibility of empty space. Bodies fall more slowly in water than in air due, apparently, to the different resistance offered by water and air. Thus, the less dense the surrounding medium, the faster is the fall of all bodies, so that in empty space bodies would fall with infinite speed, which is absurd. Hence there is *no* empty space. Space is as yet always taken to be 'filled with matter' and philosophers dared not assign any properties to absolute 'emptiness'. Democritus's materialism boldly surmounts this obstacle too: to him, matter consists of atoms separated by empty space, and geometry is a property of empty space. Other qualities too, like 'above' and 'below' are assigned to space. The acceptance of the naïve division into matter and space, without criticism, is fundamental for the progress achieved by materialism. The well known explanation of the states of matter for example, is based on this very independence of the structure of space and matter. It needs to be stressed that in this instance, too, the successes of Democritus's teachings had been achieved at the expense of an understanding of the nature of the relations of space and matter. You know that real progress in this question of 'space and matter' has only been very recently achieved in the general theory of relativity. During the whole development of science from Democritus to Newton and Maxwell, the discussion of this problem had been of no importance. Space was 'explained' by analysing its geometrical properties and by transposing the geometrical experiences of daily life, without any further thought, to the world of atoms and stars. We had done without a deeper understanding of the relation: space-matter.

In these two discussions of the concepts of matter and space we already meet the quite general problem of the real meaning of the term 'understanding' of nature. Did Democritus's atomic theory lead to an understanding of the qualities of matter or had it done without such an understanding? In what sense did the theory 'explain' the geometrical behaviour of bodies? Could the researches of Pythagoras's pupils on the oscillations of strings and their harmonies, could Democritus's suppositions be classed as 'Science'? Questions such as these had already very early been the subject of Greek philosophic thought.

You will recollect the famous analogy in Plato's *State*, in which the philosopher compares the world with a dark cave and them with prisoners, chained with their backs to the light so that they can only see the shadow of things and follow their movements. Plato describes how the prisoners take only the shadows to be real and attempt to discover the regularities of their movements. Then one of the prisoners is released and he is allowed to see the real light and things as they really are. And Plato describes how this man of real experience now thinks of his captivity and the study of the shadow images only with pity and contempt. Following this analogy the philosopher speaks about the various sciences, the knowledge of numbers, the art of measuring and the knowledge of the stars. He distinguishes four stages of perception: the highest is called ἐπιστήμη and corresponds to the knowledge of real things, the perception and recognition of their nature, as described in the analogy.[1] The second stage is called reasoned understanding—διάνοια—and can be achieved through a study of the sciences.[2] The last two stages relate to the first two as believing does to comprehending. They are called faith and belief (πίστις)[3] and conjecture (εἰκασία). In our problem, the possibility of a physical 'explanation' of nature, we are mainly concerned with the destinction between the first two stages of perception. I should like to show their approximate meaning by taking a more commonplace example. Consider a man, whom we believe we know well, suddenly committing some misdeed which is at first quite incomprehensible. Those who know all the details of the case can then explain to us the reasons for his action. Thus we are in a position to deal with all the arguments, one after the other, and eventually, after a thorough investigation of these arguments we may understand the wrong he committed. This understanding corresponds to διάνοια. Or alternatively we may suddenly realize that this man 'had to act as he did. This sort of recognition can be described by Plato's ἐπιστήμη.

Let us now return to science. Plato himself explains in great

[1] 'reason' as translated by B. Jowett in 'The Dialogues of Plato' translated into English by B. Jowett. *Republic* 511. (Clarendon Press 1875.)
[2] 'understanding', op. cit. as above.
[3] 'faith or persuasion', op. cit. as above.

detail the nature of the second, lower, stage of perception and understanding and how we can reach this level through a study of nature. The mathematical laws of nature which can be found to underlie natural phenomena appear to him of prime importance compared with the varied changes of the phenomena themselves. No other tasks of science is comparable to the search for the enduring laws of constantly changing phenomena. It is important and characteristic that Plato stresses just this aspect of science, one which we now occasionally call its 'formal' side. In one place, for example, Plato speaks about the followers of Pythagoras and their investigations into the oscillations and harmonies of strings. To him the only important feature of these experiments are the numerical relations underlying these harmonies; the phenomena themselves remain unimportant adjuncts. But the perception and understanding of nature which can be attained by the study of its mathematical structure is, in Plato's view, only the prelude to the melody proper which it is our real purpose to learn. It is a step to the knowledge of the 'nature' of things, to the first stage of true perception. Those who forget this fact are likened to the prisoners who accept the movement of the shadows as the whole of reality; they will never see the real light. The importance which the philosopher attached to a 'true' understanding of the methods of ordinary science is underlined by the position he allotted the man of full knowledge over the other prisoners, in his analogy. He writes there: 'And when he remembered his old habitation, and the wisdom of the den and his fellow-prisoners, do you not suppose that he would felicitate himself on the change and pity them? . . . And if they were in the habit of conferring honours on those who were quickest to observe and remember and foretell which of the shadows, when they moved, went before, and which followed after, and which were together, do you think that he would care for such honours and glories, or envy the possessors of them?'[1]

What of the sciences, seen historically? What is their position regarding these two kinds of perceptions? We know from history that their whole development, from Thales to the present day,

[1] 'The Dialogues of Plato. *Republic* 516 c and d, p. 402, op. cit.

has constantly augmented our 'insight into nature' ('διάνοια'). However, a contemplation of this development raises the impression that the two kinds of perception ἐπιστήμη and διάνοια, though in a sense interdependent, nevertheless stand to one another in a mutually exclusive relationship. The more new fields are opened up by physics, chemistry and astronomy, the more we are in the habit of replacing the words 'interpretation of nature' (*Naturerklärung*) by the more modest expression 'description of nature' (*Naturbeschreibung*). It becomes more and more clear that we are dealing, in this progress, not with immediate and direct knowledge but with analytical understanding. Every great discovery—and this can be seen especially in modern physics—moderates the pretensions of scientists to an understanding of the universe in the original sense. We believe that this process is deeply founded in its own nature or in the nature of human thought itself. Naturally, every attempt to show the compulsory nature of this development by means of an epistemological analysis (*erkenntnistheoretische Analyse*) of the word 'understanding', is bound to leave a feeling of insufficiency. However, this is not the place to argue the value or the necessity of this development; it seems to me to be more correct to demonstrate by means of the history of physics, including its most recent developments, how straight and consistent has been the path of science in the course of the centuries. This may convey to you the feeling of the peculiar, quite impersonal compulsion which seems to find expression in this development.

The starting point of Galileo's physics is abstract and lies exactly on the line which Plato had already mapped for science. While Aristotle had still described the real movements of bodies in nature and hence had, for example, postulated that light bodies generally fall more slowly than heavy ones, Galileo was concerned with an altogether different question: how *would* bodies fall if there were no air resistance? How will bodies fall in empty space? He succeeded in formulating mathematically the laws of this theoretical movement, though it can be only approximately realized by experiment. In place of a direct concern with the processes of nature as it surrounds us, he is concerned with the mathematical formulation of a limit which can only be checked under extreme conditions. The possibility of formulating laws from natural processes in a precise and simple manner is achieved at the sacrifice of applying these laws immediately and directly to natural events.

Copernicus's famous discovery is a step in the same direction. In order to formulate the movements of the sun and the planets in a more simple and unified way, the central position of the earth is no longer accepted as axiomatic.

This part of the development is finally and consistently concluded by the genius of Newton who formally unites into *one* law, two entirely separate fields of experience, i.e. the movements of the stars in the sky and the gravity of bodies on earth. We can, to-day, hardly conceive what an extraordinary experience it must have been for the scientists of those times to recognize that the movements of stars and the movements of bodies on earth can be traced back to one and the same simple system of laws. He who has not himself experienced the importance of this marvel can never hope to understand anything of the spirit of modern science. Yet the question must be put again: how far have Newton's discoveries 'explained' the movements of the stars? Do we really understand them better than before? To elucidate this question further it may be useful to compare the description of planetary movement in Greek science with a corresponding description in a modern textbook of Astronomy. In Plato's *Timaeus* we read in Cosmogony:

'Now, when all the stars that were needed to make time had attained a motion suitable to them, and their bodies fastened by vital chains, had become living creatures and learnt their appointed task, . . . they revolved, some in a larger and some in a lesser orbit—those which had the lesser orbit revolving faster, and those which had the larger moving more slowly.'[1]

The corresponding passage in the text-book of Astronomy by Newcomb-Engelmann reads: 'The planets move round the sun. Hence they must obey a force directed towards the sun. This

[1] 'The Dialogues of Plato'. *Timaeus* 38–9, op. cit.

force can be nothing but gravitation, the attraction of the sun itself ... Using Kepler's third law, a simple calculation shows that the force with which the planets gravitate towards the sun is in an inverse ratio to the square of their mean distances from the sun ... The question now remains: what type of curve round the sun a planet will describe under the action of such a force. Newton proved that in general the curve must be a conic section with the sun in one of the foci. Thus the mystery of the heavenly movements was resolved and the planets simply proved to be heavy bodies moving according to the same laws as we see acting around us.' This modern description is different from the older one mainly in three characteristic aspects. It puts a quantitative in place of a qualitative statement, it traces different types of phenomena back to the same origin, and it no longer considers the question of the 'why'. With regard to this sacrifice, it is characteristic that, for example, the scientists of the romantic period were not satisfied with Newton's theory, and that a distinguished scientist like Lorenz Oken attempted to replace this theory by one more 'living'. Oken wrote at one time: 'Not by mechanical manipulation (Stossen und Schlagen) but by infusing life do you create the world. Were the planet dead, it could not be attracted by the sun.'

Let us now change from Mechanics to Optics. Newton dispersed light, uniformly white to our perception, into a spectrum of different colours. Huygens replaced light by wave motions of a hypothetical medium called ether. Finally Maxwell interprets this wave motion as oscillation of the electric and magnetic fields in empty space. Here too, we clearly see how science sacrifices more and more the possibility of making 'living' the phenomena immediately perceptible by our senses, but only lays bare the mathematical, formal nucleus of the process. The fact that electric, magnetic and optic phenomena are related and that they can be traced back to the same simple system of Maxwell's equations, is no doubt a discovery of the utmost importance. On the other hand, we must admit that a blind man may learn and understand the whole of optics and yet he will have not the faintest knowledge of real light. This sacrifice of the

HISTORY OF THE PHYSICAL INTERPRETATION OF NATURE

living and immediate understanding, which had been the basis of scientific progress since Newton, was also the real reason for Goethe's bitter struggle against Newton's physical optics and his teachings on colour. It would be superficial to neglect this struggle as unimportant, there is a good reason for one of the most eminent of men using all his power to combat the achievements of Newton's optics. One can only charge Goethe with a lack of consistency. He should not only have combated Newton's views but he should have said that the whole of Newton's physics, optics, mechanics and gravitational theory was the work of the devil. It is, on the other hand, a clear sign of the strength and inner consistency of abstract science, that, in spite of all these objections, it steadily progresses in the same direction. Indeed the fact cannot be neglected that this strength is partly due to the possibility of controlling technical development with the aid of abstract science.

The rounding off of mechanics by Newton, of electricity and optics by Maxwell, and the great developments in chemistry at the beginning of the last century, directed our attention again to the problem of 'matter'. They stimulated a new desire to solve the problem whose solution the Greeks had initiated with the newly gained tools of modern science. Democritus's atomic theory was revived. Gassendi had endangered his life as early as the seventeenth century through his public teaching of atomic conceptions. His successors 'explained' the different states of matter by the supposition that the atoms are in a strict order in the solid, that they move at random but are tightly packed in the liquid and that they flit about like a swarm of midges with considerable interatomic distances, in the gaseous state. Thus the qualities density, shape and mobility were reduced to geometric configurations of the atoms. To these qualities there was added in the last century that of temperature. Heat, which had hitherto been regarded by many as a distinct substance, consisting of Democritus's atoms of fire, was now conceived as the mechanical energy of physical atoms. The movement of atoms in a hot body is faster than that in a cold one, or a strong movement of atoms causes the sensation 'warm'. As you know, all phenomena

HISTORY OF THE PHYSICAL INTERPRETATION OF NATURE

relating to heating and cooling can, on the basis of this assumption, be treated quantitatively and the theory of chemical reactions can easily be fitted into such a scheme. The qualitative changes of substances during chemical processes can apparently be traced back to changes in the geometric configurations of atoms. The processes of electrolysis show further that there are atoms of electricity, protons and electrons, and the study of radio-activity shows that these atoms of electricity must be regarded as the fundamental particles of which all the other atoms are built. Thus only protons and electrons can properly be called atoms i.e. indivisible particles, since the other so-called atoms consist of them. As you know, atomic physics, as developed under the leadership of Bohr during the past twenty years and as yet incomplete, includes an extraordinarily wide range of experiences. It contains, for example, the formal mathematical quintessence of all chemical regularities. We can, to mention only one result, calculate from it, in principle, the colours of all simple substances. Democritus's programme has thus largely been realized, the visible qualities of matter are traced back to the configuration properties of the atoms.

But modern atomic physics goes considerably beyond that of the Greeks in one point on which understanding is essential for its whole development. According to Democritus, atoms had lost the qualities like colour, taste, etc., they only occupied space, but geometrical assertions about atoms were admissible and required no further analysis. In modern physics, atoms lose this last property, they possess geometrical qualities in no higher degree than colour, taste, etc. The atom of modern physics can only be symbolized by a partial differential equation in an abstract multidimensional space. Only the experiment of an observer forces the atom to indicate a position, a colour and a quantity of heat. *All* the qualities of the atom of modern physics are derived, it has no *immediate and direct* physical properties at all, i.e. every type of visual conception we might wish to design is, *eo ipso*, faulty. An understanding of 'the first order' is, I would almost say by definition, impossible for the world of atoms. This development seems to us entirely consistent. Firstly

38

it re-establishes the balance between the various properties of matter which had been lost in the old atomic theories, the geometrical properties are no longer favoured above others. As Bohr has stressed, it is no longer correct to say that the qualities of bodies have been reduced to the geometry of atoms. On the contrary, the knowledge of the colour of a body is only made possible at the expense of the knowledge of the atomic and electronic movements within this body. Conversely, a knowledge of the electronic movement enforces the sacrifice of the knowledge of colour, energy, and temperature. Both these can only be reduced to the mathematics of the atom. In modern atomic theory, no property of bodies affecting the senses is accepted without its being analysed, nor is it automatically transferred to the smallest particles of matter. Rather is every property analysed for the purpose of διάνοια. Hence it follows as a natural corollary that atoms can have none of these properties in the usual sense.

The discussion of Newton's mechanics and optics will have already given you the feeling that the strength of this abstract development of science lies, in the first place, in its capacity to encompass large fields of experience in a simple manner and continuously to simplify and unify the picture of nature drawn by science. Atomic physics has, as is shown more clearly than ever by the progress of recent years, led to the most brilliant successes. We cannot, without admiration, pass by the fact that the infinitely diverse phenomena of nature, on earth and on the stars, can be classified by so simple a scheme of laws. On the other hand we must not forget that a high price had to be paid for this unification of the scientific concept of the universe. Progress in science has been bought at the expense of the possibility of making the phenomena of nature immediately and directly comprehensible to our way of thought.

Thus I return to the question posed at the outset: can science claim to lead to an understanding of nature? I have attempted to show how physics and chemistry—driven, we hardly know by what force—have continuously developed in the direction of a mathematical analysis of nature under the guiding principle of unification. The claims of our science to an understanding of nature, in the original sense of the word, have at the same time decreased. The attempt to prove impossible a perception-theoretical understanding of the latter kind, and to prove mathematical analysis the only possible way, appears to me as unwise as the opposite assertion, that an understanding of nature can be achieved in a philosophical way without a knowledge of its formal laws. The decision as to whether a certain type of understanding of nature can be considered satisfying and sufficient must eventually be left to the conscience of the individual or that of a period. One thing, however, modern science can certainly claim. It has, on the path of human progress, created new forms of thought and new freedoms which no other field of human endeavour could have achieved. These will be useful and important tools in every other field of work. Science has given an important example of the fact that an extraordinary extension of the most abstract fundamentals of our thought is possible without our having to accept any lack of clarity or precision.

3

Questions of Principle in Modern Physics[1]

In speaking to you on some questions of principle in modern physics, I do not wish to present you simply with a survey of the content of physics as it has developed during the last thirty years. I am certain that the strange revision of the fundamentals of exact science, forced upon us by the experimental experiences of the last decades, based on the use of more delicate apparatus, has been amply discussed here.

I would prefer to pose immediately the question: How has such a revision of our fundamental physical concepts been possible? And in view of this revision what is the 'truth-content' of classical and modern physics respectively?

In putting the question in this way we are approaching that complex of problems which—starting from the fundamental premises of quantum theory—was posed and seriously discussed by Bohr. It is not so much a real theory of perception in science as a consciousness of the fundamental laws on which the whole structure of modern physics rests.

Classical physics is based on a system of mathematically concise axioms whose physical content is fixed by the choice of the words used in them. Hence they determine unequivocally the application of these systems of axioms to nature. The validity of classical physics—like any other mathematical statement—thus appears absolute. The pronouncements of classical physics are precise and determining.

Wherever concepts like mass, velocity and force can be applied unhesitatingly, there Newton's law, that the force is

[1] Lecture, delivered at Vienna University on November 27, 1935.

equal to the product of mass and acceleration, will be shown to be true. This represents the validity of Newton's mechanics. How far this claim to validity is justified can best be seen from the fact that Archimedes' laws of the simple lever still form to-day the theoretical basis of all load-raising machines and there can be no doubt that they will do so for all time. In spite of this there has arisen in modern physics the necessity for a revision of classical mechanics. To understand this, one must examine more closely the nature of this revision. When one considers the basis of modern physics, one finds that it really does not infringe on the validity of classical physics. Rather has the necessity, and indeed the possibility, of a revision been raised by the limits encountered in the application of the system of concepts of classical physics. It is not the validity but only the applicability of the classical laws which is restricted by modern physics. The experiences which provide the basis of relativity theory have shown for example that the simple time-concept of Newton's mechanics ceases to be of use when we are dealing with bodies moving with a speed that approaches the velocity of light. It is impossible to conceive of a watch which would measure the quantity t in Newton's equations. It is for that reason that Newton's mechanics cannot be applied in this case. Again, to use an example from nuclear physics illustrating the positive side of the statement. As far as the track of an electron in a Wilson cloud chamber can be investigated, the laws of classical mechanics can be applied to it. Classical mechanics does predict the correct track of the electron. But if, without observation of its track, the electron is reflected at a diffraction grating, the basis for an unambiguous application of the space-velocity concept has disappeared and classical laws cannot be applied to such a process.

This situation shows clearly that the possibility of a revision of the exact laws of classical physics arises as a result of the lack of precision of the concepts used in the system. Thus, while the quantities x, t and M used in Newton's mechanics are linked without ambiguity by a system of equations whose solutions contain no degree of freedom apart from the initial conditions, nevertheless, the words 'space, time, mass' which are attributed to these quantities are tainted with all the lack of precision to which we have to acquiesce in everyday life. It is true that it is one of the basic experiences conditioning our science that to a certain extent communication with other people can be achieved with the aid of these words. But this again is only possible through an exact analysis of the validity of these concepts. And this in turn could only be carried out if there existed a simpler system of concepts which we could, so to speak, trust implicitly. Thus the validity of classical physics is limited by the lack of precision of the concepts contained in its axioms.

After what has just been said, it can be seen that science obviously runs the danger of being forced into a revision of its basis as soon as it leaves the field of common experience. The current concepts will lose their value for the orderly presentation of new experiences. It seems that one might escape this danger, from the outset, in science, by applying all concepts only within the limitations on which they are founded on experience; i.e. modern science should be preceded by a purification of language eliminating all ambiguous terms and concepts. But such a programme could never be carried through. The most common terms would need revision and there is no knowing how much of our language would remain. Also there is no criterion allowing an *a priori* assessment, as to whether the application of a term is objectionable or not. Before the experiences of quantum theory the results of Wilson's cloud chamber experiments could unhesitatingly be expressed in these words: 'We see in the cloud chamber that the electron has described this or that path.' Indeed we could accept this as a simple description of experimental facts. It was only later that we came to know, from other experiments, the problematic nature of the term 'path of the electron'. Therefore the only possible progress for science seemed to lie in the unhesitating use, in the first place, of existing terms for the description of experience, and the revision of these terms from time to time as demanded by new experiences. To demand a previous clarification would be equivalent to an anticipation by logical analysis of the whole of the future development of science. It is obvious, then, that the lack of precision contained in the systems of concepts of classical physics is a necessity. Hence we must also become reconciled to the idea that even the mathematically exact sections of physics represent, so to speak, only tentative efforts to find our way among a wealth of phenomena. This will obviously apply to modern as well as to classical physics. For, if certain ambiguities of the time concept have been remedied by relativity theory and certain ambiguities of the concept of matter by quantum theory, yet there can be no doubt that the future development of science will force further revisions and that the concepts used at present will also prove to be limited in their application but in a sense as yet unknown.

Here we can suitably ask the question: how can we speak of exact science at all? As an answer we can again quote an example of the range of validity of classical mechanics. So far as the concepts space, velocity, mass, etc., can be applied unhesitatingly—and that certainly applies to all experiences of everyday life—Newton's principles certainly apply. These laws therefore represent an idealization, achieved by taking into account only those parts of experience which can be 'ordered' by the concepts space, time, etc. Seen from this point of view, the forming of concepts in classical mechanics appears only a consistent extension of language. Here too, every single term represents an unconscious attempt to introduce order and communication into certain experiences by stressing common trends and by introducing a suitable notation. And just as further development of language is only possible on the basis of already existing words and terms, so in physics the concepts of classical physics form the necessary prerequisites for the investigation of atomic phenomena. Looking at classical physics as a whole then, its essential idealization consists in its ordering of experience on the assumption of objective events in time and space. Classical physics represents, in a sense, the clearest expression of the concept of matter (*Dingbegriff*), in that it attempts to make the description of the world as independent as possible of our subjective experiences. Because of this, the concepts of classical physics will always remain the basis for any exact and objective science. Because we demand of the results of science that they can be objectively proved (i.e. by measurements, registered on suitable apparatus), we are forced to express these results in the language of classical physics. Thus, for example, for an understanding of relativity theory it is essential to stress that the validity of Euclidian geometry is presupposed in the very instruments—used for the measurement of the deviation of sunlight—which are to show the variations from this same Euclidian geometry. It can also be shown, as Dingler, for example, has stressed, that the very methods used in the manufacture of these instruments enforce the validity of Euclid's geometry for these instruments (within the range of their accuracy). In a similar manner, we must be able to speak without hesitation of objective events in time and space in any discussion of experiments in atomic physics. Instructive examples of this are the experiments where the presence of neutrons is shown by the artificial radio-activity caused by them. The physical processes underlying these experiments can, without doubt, only be understood by using the abstract concepts of quantum theory. Yet the experiments are suitable for measurement because their results can be expressed in classical terms without paying attention to the abstract character of the 'quantum-theoretical' connection. Thus: 'By means of artificial radio-activity we can state that a neutron (i.e. a certain particle [*bestimmtes Ding*]) was found at that definite place at that time.'

Thus, while the *laws* of classical physics, seen from the point of view of modern physics, appear only limiting cases of more general and abstract connections, the *concepts* associated with these laws remain an indispensable part of the language of science, without which it would not be possible even to speak of scientific results.

Before the discovery of relativity theory this fact formed probably the main reason for the belief that classical concepts would have to be the constituent parts of every physical theory for all time. And even to-day, criticism of relativity and quantum theory (erroneous criticism, I believe), is based on this score.

QUESTIONS OF PRINCIPLE IN MODERN PHYSICS

Thus, it is said: it is impossible to make time relative since the discussion of every measurement presupposes absolute time. Or, in the case of quantum theory, that the use of statistical laws must always remain unsatisfactory in a description of nature. Also, that the inability to predict an event can only be looked upon as a sign of a problem as yet unsolved. Hence the question needs to be asked: How does modern physics gain the freedom to pass beyond the limits of classical concepts?

It was the increased range of technical experience which first forced us to leave the limits of classical concepts. These concepts no longer fitted nature as we had come to know it. We observed the track of an electron moving as a particle in a Wilson chamber and, on another occasion, we found it reflected on a diffraction grating like a wave. The language of classical physics was no longer capable of expressing these two observations as effects of a single entity. We had, first of all, to define more closely those places where classical concepts became ambiguous in their application.

It is the definition of the precise point at which a development beyond classical concepts has become logically possible, which represents the core of any modern theory. Thus the core of the special theory of relativity is the statement that the simultaneity of two events at different places is a problematical concept. Similarly, in quantum theory, it is of the greatest importance that to speak simultaneously of a definite position and a definite impulse of a particle is meaningless. The same statements have occasionally also been put in this way: The question of a 'real simultaneity' of two events is a 'false' problem as is the question of the exact position and exact impulse of a particle. These are questions to which there is no answer because they are put in a false way. Indeed this formulation contains the logical quintessence of the situation confronting us. It expresses in the clearest manner that the concepts, which we are forced to use in expressing our experiences, are too ambiguous to account fully for the facts of nature. What is decisive, however, is not the statement, that there *are* 'false' problems, but a reason *why* they exist.

The special theory of relativity states that there is, up to the present, no means of transmitting signals with a velocity greater than that of light. Hence it is impossible to give a clear definition of an absolute time-scale. This, however, is a negative statement. Only the supposition that it is *in principle* impossible to transmit signals with a speed faster than light, and arising from this the postulate of the constancy of the velocity of light, makes possible a logically satisfying ordering of experience. It is only this second positive step that justifies the statement that the question of an absolute time-scale is a 'false' question. The same applies to quantum theory. The restrictions of classical concepts as enunciated in the uncertainty relations acquire their creative value only by making them questions of principle. They then afford the freedom necessary for a harmonious and non-contradictory ordering of our experience. Only the system of mathematical axioms of wave and quantum mechanics entitles us to class the question of position and impulse values as a 'false' problem.

The appreciation of the logical situation in which an apparently correctly formulated question becomes devoid of meaning, has thus become the precondition for an understanding of modern physics. On the other hand, modern physics also shows that the relegation of a question to being a 'false' problem is only possible and can only become fruitful on this condition: it must create the freedom necessary for the establishment of the required abstract interconnections. In our approach to a description of nature we use concepts which lack precision in certain respects, though we naturally cannot appreciate that at the time. Yet finding these weaknesses will lead to new knowledge only if they can be used in a definite way for an appreciation of new kinds of interconnections. So long as this has not been done, we have no reliable criterion for asking whether a problem has or has not a meaning. We must rest content with treating all theses of physics—even those formulated mathematically—merely as word images, since we cannot know the range of accuracy of the terms and concepts used. We are merely endeavouring to make our experience of nature intelligible to ourselves and to others.

However, once these new connections have been established we can penetrate into a new world of concepts qualitatively different from the old. In this way relativity and quantum theory represent the first decisive step out of the field of apprehensible concepts into an abstract field, as yet untouched, and the character of the connections discovered in it leave no doubt that these steps can never be retraced. Of course, these new connections cannot claim to use concepts better defined than the classical ones and they may yet have to be revised in the future. Nevertheless, the concepts developed in these theories have proved themselves to such an extent in the ordering of the more delicate experiences, that we have reason to believe them as suitable for our new experiences as the old concepts were for the experiences of everyday life. Hence they will in their turn become the precondition for any further development of physics. After all, the discovery of a new system of concepts means nothing more than a new method of thought which can never be annulled as such.

For this reason the real situation in our science can in no way sustain the hope, occasionally expressed, that at some future date classical concepts may yet be used for the ordering of relativistic and atomic phenomena. It is more likely that there is a certain range of experience which can be interpreted by Schrödinger's wave mechanics but not by classical mechanics, and we must assume that even the less palatable features of the laws (Gesetzmässigkeiten) of quantum mechanics will remain integral parts of theoretical science. As an example, I should like to discuss the finality of the statistical character of quantum mechanics and whether any hope can be entertained of extending and completing quantum mechanics on a determinist basis. Indeed there could apparently be no objection to an assumption that, say, the radium atom possesses hitherto unknown properties which accurately define the time of emission and the direction of an α-particle. However, a more detailed analysis shows that such an assumption would force us to consider as wrong those very statements of quantum theory which allow an accurate mathematical prediction of experimental results. We have, so far, had every reason to rely on those parts of quantum mechanics. I should like to deal with this in greater detail.

Any experiment in atomic physics starts with the following situation. With the aid of more or less complicated apparatus we put questions to nature directed towards establishing some objective process in space and time. We may, for example, want to know whether electrons are deflected at a certain place. In this situation it follows automatically that, in a mathematical treatment of the process, a dividing line has to be drawn between, on the one hand, the apparatus which we use as an aid in putting the question and thus, in a way, treat as part of ourselves, and, on the other hand, the physical systems we wish to investigate. The latter we represent mathematically as a wave function. This function, according to quantum theory, consists of a differential equation which determines any future state from the present state of the function. But we are satisfied with the laws formulated in terms of classical concepts for the making of our apparatus and feel entitled to use them for measuring purposes. The dividing line between the system to be observed and the measuring apparatus is immediately defined by the nature of the problem but it obviously signifies no discontinuity of the physical process. For this reason there must, within certain limits, exist complete freedom in choosing the 'position' of the dividing line. The behaviour of the measuring apparatus must not, of course, contradict the laws of quantum mechanics. Indeed quantum mechanics contains the laws of classical mechanics as a limiting case and the position of the dividing line can be freely chosen within certain limits. The laws of quantum mechanics assume their statistical character only at the dividing line, because the physical connections, on both sides of the dividing line, can be unambiguously formulated. The possibility of statistical inter-connections is created only by regarding the effect of the measuring apparatus on the system to be measured as a partial disturbance uncontrollable in principle. Thus the only place for a determinist supplement to quantum mechanics would be at the dividing line. Since, however, the new physical properties to be determined must be attributed to a definite system, a contradiction between the logical (*gesetzmässigen*) consequences of the new properties and the connections of quantum theory becomes inevitable, as soon as the dividing line is removed from the system. For the new physical properties of the observed system, which were supposed to close the gaps in the statistical laws, would now, after removal of the dividing line, appear in a position, where there is no room whatsoever for a supplement: They could only disturb the already existing unambiguously determined (*gesetzmässigen*) connections.

This chain of reasoning is particularly applicable in the case of radio-active disintegration. The α-particles, emitted by a nucleus, are reflected at a diffraction grating, according to their accurately known energies, in clearly defined directions; these directions are determined by the properties of the *whole* grating. Now, if there existed such an unknown property of the radium atom which would allow us to predict in which direction the α-particle was emitted, then we could also predict on which part of the diffraction grating the impact will take place. Thus the direction of reflection could not be determined by the whole grating, and a contradiction arises. This contradiction is really the result of our classical interpretation of the statement: 'an α-particle moves along a certain track.' Hence we assume that 'its reflection cannot depend on the nature of the grating at some distance'. But without this interpretation we could not decide what is meant by the statement 'the α-particle moves just at *this particular* place'. In the last resort, we have to fall back on unquestioning application of classical concepts at some point, if not on the α-particle, then on the apparatus designed for its observation.

It should also be mentioned in this connection that the statistical character of quantum theory is in many respects fundamentally different from that used in the kinematic interpretation of thermodynamics. The degree of accuracy of the latter theory always expresses our lack of knowledge of the system concerned. But in quantum theory, ignorance of the result of future experiments *can be* compatible with the complete understanding—in the usually accepted sense—of the state of the system concerned. For instance, the statement that an atom is in its normal state implies a complete knowledge of the atom concerned. This can be shown, because we can, on the basis of this knowledge, derive the mutual reactions of this atom and any other system, and also because there are certain experiments whose results we *can* accurately predict. There are, of course, other experiments for which accurate prediction is impossible. As I explained earlier, the *definite* statements of quantum mechanics indirectly result in making it impossible to supplement the statistical statements. On the other hand in Heat, ignorance about the results of certain experiments is always identified with ignorance of the true state of the system and is shown up in all experiments. Therefore in classical statistical mechanics, uncertainty about the result of a future experiment can be taken to indicate an as yet unsolved problem. But this does not apply to quantum theory since quantum theory always enables us to give full reasons for the occurrence of an event after it has actually taken place.

Finally I want to raise the question of those points at which modern physics itself will have to be revised. It is obvious that the range of application of the new concepts, too, will necessarily be limited. As a result of the discoveries of the last few years, it has become probable that the existence of the electron will enforce the next limitations in the application of present concepts. The existence of the electron is closely connected with the problem of harmonizing the demands of relativity with those of quantum theory. This is made clear by means of the dimensionless constant $\frac{e^2}{\hbar c}$ (Sommerfeld's constant). But the problem cannot be tackled without accepting, much more than has been the practice, that matter and radiation are different effects of a unified (einheitlich) event. A first step in this direction was the discovery of the possibility of transforming matter into radiation and vice versa (Dirac and Anderson). Through these discoveries a series of new problems has been raised concerning the measurement of fields, positions of electrons, etc. Eventually an understanding of the nature of the electron may make it imperative to take into account the atomistic structure of all measuring apparatus, a new step not required in quantum mechanics. From previous experience there can be little doubt that the new theories will differ from previous quantum mechanics again by branding certain questions—at present considered sensible—as 'false' problems. In spite of this, it must be stressed again that any such new development will have many aspects whose meaning may not at first be at all clear. A typical example of this is Dirac's theory of 'holes'. It would be difficult to assign any meaning to the statement that the whole world is filled with electrons of negative energy at infinite density. Nevertheless this formulation of electronic theory has proved so useful that it was not only possible to predict the existence of the positron and its break up by radiation (Zerstrahlung), but also to derive from it important modifications of Maxwell's theory of a large and fast changing field. The possibilities arising from this have not been by any means exhausted.

Dirac's theory shows here the really fundamental characteristic of a physical discovery. It is not the result of, but the precondition for a clear delineation of the range of applicability of the discovered concepts. The theory must open up new methods of thought and hence effect a real change in the theoretical situation and force us to alter the way in which we put our questions to nature. In other words, it must lead to a new harmony—not hitherto achieved—in the field to which it applies.

In conclusion, I may be allowed to say that it should not be taken as scepticism if I suspect that the concepts of modern physics too will have to be revised. On the contrary, it is only another expression of the conviction that our ever-extending field of experience will bring to light ever more new harmonies.

4

Ideas of the Natural Philosophy of Ancient Times in Modern Physics[1]

Modern science has followed many trends of early Greek natural philosophy by reconsidering the problems with which that philosophy had grappled in a first attempt to understand the surrounding world. Hence it may be well worth considering which of those early ideas have retained their creative power in modern physics, and what shape they have acquired by absorbing the scientific experiences of the intervening two thousand years. There are, especially, two ideas of early Greek philosophy which to-day still determine the course of science, and which are therefore of special interest to us: the conviction that matter consists of minute indivisible units, the atoms, and the belief in the purposely directive power of mathematical structures.

The thesis of the existence of atoms was the natural consequence of the development of the concept of matter, the classification of which was the first endeavour of ancient natural philosophy. The conviction that, in the transience of phenomena, there must be something permanent which is subjected to change, led to the teaching of the existence of some 'fundamental matter'. For Thales, this fundamental substance was simply water, on which all life appeared to depend. His successors defined this concept more accurately and attributed to it the characteristics of entity (Einheitlichkeit) and indestructibility. Thus, to make intelligible the variety of phenomena, several

[1] First published in *Die Antike* (Organ der Gesellschaft für Antike Kultur.) Vol. XIII.

kinds of 'fundamental matter' had later to be postulated whose mixture and separation is the cause of the manifold changes of events, unless of course these permanent elements were to be something beyond the material world. Thus Earth, Fire, Air and Water appeared to be the natural elements of which the world was constructed. To make these ideas into a real explanation of phenomena, the process of mixing had to be clearly described. It seemed obvious to interpret the mixing of two liquids essentially like that of a mixture of water and sand, and to assume that the smallest particle of the liquid would retain its initial qualities unchanged and that these particles are present in the mixture in a random distribution. Thus originated, as if of its own accord, the idea of the smallest indivisible unit of matter and in the teachings of Leucippus and Democritus the 'atoms' appeared as the real carriers of material and spiritual evolution.

According to this view, the atoms differed no longer through their inherent qualities, but only through shape, position and movement. The protagonists of these 'atomic' ideas believed that such geometrical properties would suffice to explain all the diversity of natural phenomena. The atoms were the essential reality, between them was 'nothing', empty space. Larger composite bodies were formed by the compounding of like atoms, and their properties in turn were determined by the type of arrangement. The atoms themselves were eternal and indestructible. We shall now compare modern atomic theory with the corresponding ideas of antiquity, based on the principles I have outlined.

Modern atomic theory, too, assumes indivisible elementary particles of matter called 'electrons, neutrons and protons', and it, too, endeavours to trace back all perceptible qualities of substances to the dynamics of the atoms. However, the necessity of explaining the most delicately executed experiments to the last detail, has made it obvious that there was in ancient atomic theory a peculiar contradiction, an inner inconsistency. The basic idea of atomic theory had to be carried to its logical conclusion. Democritus's atomic theory, on the one hand, realizes that it is impossible to explain rationally the perceptible qualities of matter except by tracing these back to the behaviour of entities who themselves no longer possess these qualities. If atoms are really to explain the origin of colour and smell of visible material bodies, then they cannot possess properties like colour and smell. Thus ancient atomic theory consistently denies the atom any such perceptible qualities. On the other hand they are left with the quality of taking up space so that one may speak of position, arrangement and size of atoms. Democritus, here, patently goes beyond the opinion of his predecessors. The fundamental concept of previous philosophy, that of *being* (*Seiend*) and *non-being* (*Nichtseiend*) he brings down to earth by making it *full* and *empty*. To him, empty space is a reasonable (*sinnvoll*) concept. He thus creates the possibility of explaining the different perceptible qualities of substances by means of variable arrangements of atoms in space. But, at the same time, he has to give up the idea of linking space and time with the existence of matter, to *explain* space and time. The old and great idea that space and time are, so to speak, stretched out by matter and in essence akin to it, has no room in Democritus's teachings.

Modern atomic theory shares *this* fundamental thought with its ancient counterpart: It tries to explain the qualitative variety of external physical events by relating it to a variety of forms, which can be surveyed and analysed. Of such forms the Greek philosophers had at their disposal only the geometrical configurations, and hence ancient atomic theory explained qualities by varying groupings of the atoms in space. The preference for one particular perceptible quality—the taking up of space as a quality of atoms—does however seem to show a lack of consistency, and it is obvious that modern theory will have to differ fundamentally at this point. The indivisible elementary particle of modern physics possesses the quality of taking up space in no higher measure than other properties, say colour and strength of material. In its essence, it is not a material particle in space and time but, in a way, only a symbol on whose introduction the laws of nature assume an especially simple form. Modern atomic theory is thus essentially different from that of antiquity in that it no longer allows any reinterpretation or elaboration to make it fit into a naïve materialistic concept of the universe. For atoms are no longer material bodies in the proper sense of the word, and we are probably justified in claiming that in this respect modern theory embodies the principal and basic idea of atomic theory in a purer form than did ancient theory. Without going into detail it is naturally difficult to convey a picture of the place the atom occupies in modern science, and of the mathematical forms whose variety present us with a picture, faithful to the last detail, of the diversity of phenomena. A parallel may, perhaps, illuminate the symbolic character of the present-day concept of the atom. The atom of modern physics shows a distant formal similarity to the $\sqrt{-1}$ in mathematics. Though elementary mathematics maintains that among the ordinary numbers no such square root exists, yet the most important mathematical propositions only achieve their simplest form on the introduction of this square root as a new symbol. Its justification thus rests in the propositions themselves. In a similar way the experiences of present-day physics show us that atoms do not exist as simple material objects. However, only the introduction of the concept 'atom' makes possible a simple formulation of the laws governing all physical and chemical processes.

The abstract nature of the modern concept 'atom', and those mathematical forms which, in modern theory, serve to express the imagery for the variety of atomic phenomena, do already lead us to the second fundamental principle which our science has taken over from antiquity; that is the idea of a purposeful and directive force inherent in mathematical formulations.

We meet this idea, stated clearly for the first time, in the teachings of the school of Pythagoras, expressed by their discovery of the mathematical conditions of harmony. In investigating the vibrations of strings, they found that the condition for two strings to sound in harmony (all other properties being equal) was that their length must be in simple ratio. This means that a totality of sound appears to the human ear to be in harmony only if certain simple mathematical relations are realized, though the listener may not be conscious of this. This discovery represents one of the strongest impulses of human science, and its effects, in nature as well as in art, can constantly be seen, once the creative force of mathematical order has been appreciated. I would mention the kaleidoscope as a specially simple and obvious example. Here, something beautiful and orderly arises from a random picture, through simple mathematical symmetry. More valuable and important examples can be found in an analysis of any work of art or, in nature, in the study of crystals. If the essence of a musical harmony or a form of fine art can be discovered in its mathematical structure, then the rational order of surrounding nature must have its basis in the mathematical nucleus of the laws of nature. Such a conviction found its first expression in the Pythagorean teaching of spherical harmony, in the attribution of regular shapes to the elements. Thus in *Timaeus* Plato explains the atoms of earth, fire, air and water as cube, tetrahedron, octahedron and icosahedron respectively. But in the last resort the whole of mathematical natural science is based on such a conviction.

Modern science has thus accepted from antiquity the idea of a pattern capable of mathematical description, but it carries it out in a different manner, rigorous and, we believe, determined for all time. The realm of mathematical forms at the disposal of ancient science was still comparatively limited. They were primarily geometrical forms which were related to natural phenomena. Hence Greek science searched for static patterns and relationships. The subjects of its investigations were the unchangeable orbits of the stars, or the forms of the everlasting and indestructable atom. However, the laws that could be derived from those assumptions could not accommodate the experiences of later centuries based on the use of more delicate apparatus. Modern science has demonstrated that in the real world surrounding us, it is not the geometric forms but the dynamic laws governing movement (coming into being and passing away) which are permanent. Even Kepler thought he had found in the orbits of the stars the harmonies of Pythagoras's school. Science since Newton has attempted to see them in the mathematical structure of the law of dynamics, and in the equation formulating this law.

This change does represent a consistent execution of the programme of the Pythagoreans inasmuch as the infinite variety of natural events finds here its faithful mathematical replica in the infinite number of solutions to an equation. Newton's differential equation of mechanics serves as a good example. The demand that there should arise from the one natural law, already formulated, an infinite variety of phenomena accessible to experimental investigation, does at the same time provide the guarantee for the correct formulation of the law, which is then valid for all time. An equation formulating such a law expresses in the first instance only the simplest physical circumstances: it defines the dynamic concepts necessary for an understanding of the natural phenomena concerned. Beyond that, it contains some general expressions about the world of our experiences, like the fact that in empty space direction and position cannot be defined. Yet it encompasses, as a possible development, an infinite variety of phenomena, just as a fugue of many parts can be developed from the few notes of a musical theme. Thus, while ancient philosophy attributed regular shapes to the atoms of elements, a mathematic equation must belong to the elementary particle of modern physics. This equation formulates the natural law governing the structure of matter. It embraces the progress of, say, a chemical reaction, as well as the regular shapes of crystals or the pitch of a vibrating string. It develops logically, from the accidental initial conditions, the physical phenomena of the surrounding world, like a kaleidoscope which creates an ingenious pattern from an accidental conglomeration of coloured glass.

The successes of this method have confirmed the beliefs of the Pythagoreans to an unforeseen extent. It has partly resulted in a real mastery over the forces of nature, and thus decisively intervened in the development of mankind. Hence modern science has retained confidence in a simple mathematical basis for all regular interrelations of nature, even of those which we cannot as yet grasp. Mathematical simplicity ranks as the highest

ANCIENT NATURAL PHILOSOPHY AND MODERN PHYSICS

heuristic principle in exploring the natural laws in any field opened up as a result of new experiments. In such a case the inner relations seem to be understood only when the determining laws have been formulated in a simple mathematical way.

This search for the mathematical structure of phenomena, as taken over from antiquity has, however, given rise to an accusation. It is said that it illuminates only certain and, at that, not the most essential aspects of nature and, rather than being of help in an immediate and general understanding of nature, it is actually a hindrance. This complaint can best be answered by drawing attention to the starting point of Pythagoras's teachings. It is the conscious understanding of the rational numerical relations underlying musical harmonies which make possible both the construction and use in performance of a musical instrument. It is, however, in the unconscious mental acceptance of these rational relations that we can grasp the real content of music. Similarly, the precondition for an active, practical intervention in the material world, is just this conscious knowledge of mathematically formulated natural laws. Behind this, however, there is a direct understanding of nature unconsciously accepting these mathematical structures and mentally recreating them. All human beings are capable of this understanding if they are willing to enter into a more intimate receptive relation with nature.

5

The Teachings of Goethe and Newton on Colour in the Light of Modern Physics[1]

To advance science, in co-operation or in competition with others, it is sufficient to concentrate one's whole power on the small circle of an intended project. To survey progress as a whole, however, it is useful to make repeated comparisons with the scientific tasks of a previous era and to explore that peculiar change which a great problem undergoes in the course of decades or perhaps centuries. Such a problem, posed in a creative way, can appear again and again in a new light, even though it may have been satisfactorily answered at the time.

The continuous movement of modern science towards an abstract control of nature, removed from an appreciation based on common experience, recalls immediately the memory of the great writer who, over a hundred years ago, dared to fight for a more 'living' science in the field of colour theory. That battle is over. The decision on 'right' and 'wrong' in all questions of detail has long since been taken. Goethe's colour theory has in many ways borne fruit in art, physiology and aesthetics. But victory, and hence influence on the research of the following century, has been Newton's. The extraordinary development of Newton's physics since that time has stressed the consequences of this trend in research more than ever. The cold abstract concepts which enable us to control nature, as for example in modern nuclear physics, illuminate more clearly to-day the

[1] This lecture was delivered in Budapest on May 5th, 1941 before the Society for Cultural Collaboration.

background of that famous dispute. It is primarily this background that I wish to discuss.

We know that Goethe felt impelled to concern himself with Nature during his Italian journey. The geological structure of the country, the variety of plants flourishing under a southern sky, the vivid colours of the Italian landscape, captured his interest again and again and are brought to life for us in the vivid description in his diary. We also see from the notes how these impressions, as if by themselves, assume some scientific order and how there arise from an immediate experience of nature, concepts later destined to become the foundations of Goethe's 'Contemplations of Nature'. After his return to Weimar Goethe began to make use of his freshly gained experiences. The first result of this work, 'The Metamorphosis of Plants', was published in 1790. Work on the theory of colour which Goethe had started in Italy remained for the time in the background. On his admission, it had been stimulated by the impact of Italian colouring. After his return Goethe had borrowed a prism from Hofrat Buttner in Jena to study the colour effects of refraction. It remained wrapped on his table. Probably in spring 1791 the owner asked for the return of the prism and sent a servant to collect it, and only then did Goethe make use of the opportunity to observe the well-known colour effects. He discovered to his great surprise that large white surfaces do not, as he had assumed from his studies of Newton's theories, appear coloured, but that they are white, and that a corresponding observation applies also to large dark surfaces. Coloured borders appear only at the edges of light and dark surfaces. From this Goethe realized 'that a boundary is necessary to produce colours'. This discovery, which Goethe believed to be in contradiction to Newton's theory, provided the incentive for intensive work on the origin of colour in the process of refraction. Goethe concluded that colour is created by the combination of dark and light and not by light alone as Newton thought. This conclusion he finds confirmed by many other phenomena. The sun, which appears radiantly white during the day, looks yellow and red when obscured by an intervening layer of mist. Smoke rising from a chimney assumes in sunlight a bluish haze. Further convinced by a variety of other experiences Goethe finally came to believe in the origin of colour out of *light* plus *dark* and to have found the 'underlying phenomena' ('Urphänomen') in the admixture of dimness (Trübe) with light. This concept brings together into a unified, orderly whole the many effects of colours in our world of the senses rather by way of a guiding idea, based not on reason but on experience. The harmonious arrangement laid before us by Goethe's colour theory gives even the smallest details a living content and comprises the whole range of objective and subjective colour phenomena. Just those colours which are conditioned by processes in the eye itself and are therefore really based on an 'illusion' of our senses, are treated with particular care. And when Goethe speaks of the *Urphänomen* of the origin of colour in one of the most beautiful poems in the *Westöstlicher Diwan* then we can sense the importance this discovery had assumed for him.

Goethe had believed that the contradiction between his and Newton's theories could not be resolved, so for that reason we must now deal with Newton's theory too. This theory forms to this day the basis of all Physical Optics. White light is considered to be composed of light of different colours, similar in a way to the sound of distant breakers which is composed of the rush of individual waves though appearing to our sensations as one undivided whole. By means of external influences individual colours can be singled out. In this process of separation some matter is always required which will remove light, and this can be compared with what Goethe calls *dimness* or *darkness*. Thus Newton's theory also explains that colours are created from white light only as a result of reciprocal action with *dimness*. But the order of phenomena is completely different in the two theories. The simplest phenomenon in the Newtonian theory is the narrow monochromatic ray purified by complicated mechanisms from light of other colours and directions. Goethe's simplest concept is the bright all-embracing daylight. Newton's basic effect, so remote from our daily experience, makes optical phenomena accessible to measurement and mathematical treatment. Radiation and propagation of light can be determined by measurement and fixed in mathematical form. Every colour can be associated with a number—in modern notation the wave length. This makes optics into what is commonly called an exact science, in that it enables us to construct accurate optical instruments which open up parts of the universe not normally accessible to our senses. Newton's theory makes possible a certain control over the phenomena of light and their practical use but it is plainly of no assistance to a better appreciation of the world of colour surrounding us.

This comparison shows that mutual criticism was bound to arise between Goethe's and Newton's theories. Newton's starting point appeared strange and unnatural to Goethe. White light, that is, light in its purest form, is to be downgraded to a composite. Instead the physicist is to accept as the basic form a light tormented and forced through narrow slits, lenses, prisms and all sorts of complicated devices. We can well understand when Goethe gives vent to his disappointment in these words: 'The physicist also wins mastery over natural phenomena, he accumulates experiences, fits and strings them together by artificial experiments . . . but we must meet the bold claim, that this is *Nature*, with at least a good-humoured smile and some measure of doubt. No architect has yet had the notion of passing off his palaces as mountains and woods.' In general he deprecates the desire of physicists to penetrate through the world of phenomena, as they appear, to the causes of those phenomena. 'Even if an *original* phenomenon (*Urphänomen*) were found, there remains the dilemma that it will not be recognized as such, that we shall look behind and beyond for something further. Otherwise we should admit a limit to *seeing*. Let the scientist leave *original* phenomena in their eternal peace and splendour'.

On the other hand, the physicist can legitimately reproach Goethe in that his theory cannot be regarded as scientific since it cannot lead to a real control of optical phenomena. Particular colour phenomena, not yet observed, cannot be predicted with any accuracy. This, however, is precisely what Newton's theory can lay claim to do. Goethe's theory also deliberately links certain elements whose careful separation is the constant preoccupation of the physicist. The first presupposition of all research is the separation of subjective and objective. Thus Goethe's colour theory can enrich the physicist's knowledge in particular fields. He can learn something about the reaction of the eye to the impact of colour, about the colours of chemical compounds, or about refraction phenomena. But it is just the very unity of Goethe's theory which he cannot accept. For the reactions of the eye must find their explanation in the finer biological structure of the retina and optical nerves, which conduct the colour impressions to the brain. The colours of chemical compounds must be capable of calculation from their atomic structure, and refraction phenomena are derived mathematically from the properties of a propagating wave. . . . On this basis, an immediate connection of the three phenomena appears unintelligible. We see here a general characteristic of nature. Processes appearing to our senses to be closely related often lose this relation when their causes are investigated.

It is clear to all who have worked more recently on Goethe's and Newton's theories, that nothing can be gained from an investigation of their separate rights and wrongs. It is true that a decision can be taken on all points of detail and that in the few instances, where a real contradiction exists, Newton's scientific method is superior to Goethe's intuitive power, but basically the two theories simply deal with different things. It is much more to the point to ask how it is possible to link the idea of colour with such different subjects.

It has been said that Goethe's and Newton's methods proceeded in two entirely different directions. While Newton obviously endeavoured to open the world of colour to exact measurement and thus to create order in that world by mathematical methods similar to those he had so successfully employed in mechanics, such mathematical considerations do not figure in Goethe's work. On the contrary, Goethe explicitly dispensed with all relations of his theory to mathematics though he stressed that the assistance of exact measurement may have been desirable in some instances. On closer scrutiny, however, this difference assumes much less importance than appears at first sight. Goethe does not renounce mathematics itself but rather mathematical manipulation. When we consider mathematics in its present form, as revealed, say, in the theory of symmetry and number, it is easily seen that Goethe's theory contains no small amount of mathematics. In the section 'The Sensual-Moral Effect of Colour' he deals for example with symmetrical arrangement of colours according to polar relations. He presents an arrangement of the six primary colours in a regular hexagon or a circle divided into six equal parts: red, blue-red, blue, green, yellow and orange, in this order. Each colour in this circle lies opposite its complementary, thus red opposite green and blue opposite orange. This symmetrical arrangement of the colours led him to a study of the varied relations between them. Colours lying opposite result in 'pure, self-motivated, harmonic combinations which always carry totality with them'. The combination of two colours which are separated by only one intervening colour, Goethe calls characteristic because, as he says, 'they all have something significant which forces a certain impression upon us but does not satisfy us. This is because each characteristic only originates by a separation of a part from a whole to which it is related without dissolving itself in it.' Finally, the combination of neighbouring colours he calls a 'characterless combination'. This treatment of colour relations on a colour disc immediately recalls mathematical symmetries such as are found in an artistic ornament or demonstrated in the simplest form in a kaleidoscope. Similar symmetrical arrangements can be found throughout the whole work.

A somewhat clearer picture of the differences between the two theories can be gained by enquiring into the purpose they are to serve. This should not be misunderstood to mean that a scientific theory is always related to a definite purpose and that its only aim is to achieve this purpose. But every scientific theory arises in a certain mental climate which implies some idea as to how the projected theory might later be applied. This background is often conditioned by the historical development of the science concerned and the author of the theory may be only vaguely conscious of it. If we then speak of the purpose of a theory, in this sense, there can be no doubt that Goethe's theory of colour was designed to serve the artist, particularly the painter. Goethe himself describes at length how much he has missed a theory of colour in art and how it struck him that 'artists acted merely on the basis of vague tradition and a certain impulse and that dark and light, colouring and the harmonies of colours move curiously and without rhyme or reason.' It is certain that Goethe's initial desire was to create such a theory of colour. Beyond this desire, as a more general background, there was a goal, first mentioned during his Italian journey in connection with his plans for a theory of colour. 'I can see that with some effort and persistent thought I shall be able to experience further enjoyment of this world.'

The background out of which Newton's theory grew was entirely different. The experiences of science since Galileo and Kepler have taught us that mechanics can be summarized in, and understood by, mathematical laws. Newton was the first scientist to realize to what extent such a penetration of nature was possible. In optics, too, there existed a series of investigations, showing that large parts of this subject could be mastered with the aid of laws capable of mathematical formulation. It is quite obvious that Newton's efforts were directed precisely towards such progress in the mathematical explanation of colour. It is difficult to estimate to what extent, at that time, this desire was linked with the realization that an accurate knowledge of the physical laws can lead to the technical mastery of nature. But the fact that Newton made long and detailed investigations in the improvement of telescopes tends to show that he was quite familiar with this side of science too.

Later developments have shown how well the two theories had achieved their stated objectives. Without a mathematical theory of light there would never have been a telescope or a microscope. On the other hand many painters gathered knowledge and enrichment from Goethe's theory.

It has also frequently been said that behind this diversity of purpose there lies a deeper difference of mental approach and that the fundamentally different attitudes of the poet and the mathematician to the world have led to such different theories. This certainly expresses an important reason for the dispute, but it would be unjust to conclude that this other poetic side of the world need necessarily be alien to the scientist. We need only mention Kepler who, after all, helped to create the most important foundations of this mathematical science. Kepler always sensed in all his varied and intricate speculations on number the harmony of spheres. Listening to the enthusiasm with which he celebrated new discoveries about the harmony of planetary orbits it would be ungenerous not to credit him with definite poetic sensibility. Newton devoted a large part of his life to philosophical and religious investigations and it is probably correct to say that the world of poetry has been familiar to all really great scientists. The physicist, at any rate, also seeks to discover the harmonies of natural events. On the other hand, it would be an equal mistake to believe that the poet Goethe had more interest in arousing a vivid impression of the world than in acquiring a real understanding of it. Every genuinely great work of creative writing transmits real understanding of all aspects of life otherwise difficult to grasp. This is especially true of a work like the theory of colour which must transmit new understanding and is written with full claims to scientific accuracy.

Perhaps the difference between the two theories is most accurately defined by saying that they deal with two entirely different levels of reality. We must remember that every word of our language can refer to different aspects of reality. The real meaning of words often emerges only in their context or is determined by tradition and habit. Modern science soon made a division of reality into objective and subjective. While the latter is not necessarily common to different people, objective reality is forced on us from the outside world always in the same way and for that reason early science made it the subject of its investigations. In a way, science represents the attempt to describe the world to the extent that it is independent of our thought and action. Our senses rank only as more or less imperfect aids enabling us to acquire knowledge about the objective world. It is only natural and consistent for the physicist to try and improve on the senses through artificial means of observation until we penetrate to the most remote fields of objective reality which are entirely beyond the range of our immediate perception. At this point arises the deceptive hope that further refinement of our methods of observation may eventually enable us to get to *know* the whole world.

To this objective reality, proceeding according to definite laws and binding even when appearing accidental and without purpose, there stands opposed that other reality, important and full of meaning for us. In that reality events are not counted but weighed, and past events not explained but interpreted. Useful (*sinnvoll*) interrelations here mean a 'belonging together' within the human mind. True this reality is subjective but it is no less powerful for all that. This is the reality of Goethe's theory of colour. Every type of art is concerned with this reality and every important work of art enriches us with a fresh understanding of its scope.

It appears at first sight as though an unbridgeable gulf will for ever divide these two realities. Goethe's struggle against Newton's theory would appear simply an expression of an irreconcilable conflict. The development of science in the last few decades, however, has shown that a division of the world into two sections creates a very crude image of reality. To understand this we shall have to consider more recent developments in the realm of physics.

The idea that our senses are only imperfect aids in the appreciation of the objective world has guided science further and further away from our immediate world of the senses. A more refined technique of observation has brought to light new aspects of nature previously concealed from us, while parallel to this development the concepts of science have become more abstract and remote from common experience. A basic concept of Newtonian optics, the monochromatic ray of light, is already an idea to which we are accustomed in everyday life. The movement of science away from our world of the senses becomes quite plain when considering electrical phenomena. During the first half of the last century attempts had been made to link electrical theory with mechanics through the concept of force. However, the discoveries of Faraday and Maxwell have shown that electric and magnetic phenomena can best be understood by basing them on the idea of the electric field. True, the field concept can be made plainer by comparison with the oscillations of elastic bodies but this is obviously a simile for showing mathematical interrelations, and has no connection with our immediate sense-impression of electricity. For even when we talked of an ether whose elastic oscillations had an electric effect, this ether was outside the range of our sense-impressions. At the same time, however, this science, in becoming more and more abstract, reveals a new power. It can recognize the interconnection between the most diverse phenomena and relate them back to a common root. It is the finest justification of our enquiry into the objective world that it has led to unexpectedly wide interconnections, and that, in spite of all the complexity of detail, it has, more and more, simplified our ideas of nature. Through Maxwell's discovery, light was recognized as an electromagnetic phenomenon. This led in turn to the recognition that electric and magnetic effects, light, invisible ultra violet and infra red rays and heat radiation are but different aspects of the same physical effect in spite of the fact that they belong to entirely different parts of our world of the senses. This development is carried to its logical conclusion in modern atomic physics. Atomic physics undertakes to explain all properties of matter accessible to our senses of our experiments, by tracing them back to properties of the atom. These latter can be laid down in simple mathematical laws. Thus the infinite variety of phenomena is reflected in the infinite number of deductions from a simple system of mathematical axioms. In fact modern atomic physics can explain, from the properties of atoms, the properties of solids, chemical regularities, the effects of heat and anything else arising from an observation of matter. It is true that up to the present this explanation has been carried out, with the precision ultimately required, only in relatively few cases, but in all these cases our theory has stood up to the most rigorous tests in a wonderful way. But from an explanation of the *sense properties* of matter, from their atoms, it becomes clear that no such sense properties can be attributed to the ultimate 'brick' of matter in a simple way. While the atom can be observed, in its effects, by an extraordinary refinement of experimental techniques, it is no longer subject to our immediate sense perception. The scientist must reconcile himself to the idea of directly linking the fundamental concepts on which his science rests with the world of the senses. They justify themselves as fundamental concepts because they penetrate the infinite variety of phenomena of our world of the senses, introduce regularity and order and thus make it comprehensible. This is proved by the technical developments they have made possible and which enable man to harness the forces of nature to his purpose.

This development has been responsible for a peculiar change in our views concerning the objective world of science. Our intention to eliminate those errors which may have been introduced by the deceptions and inaccuracies of our perception, caused us to describe the world in a manner entirely independent of our own thoughts and actions. The idea was to sketch as accurate a picture of nature as possible. Now it has turned out that this picture becomes, with increasing accuracy, further and further removed from 'living' nature. Science no longer deals with the world of direct experience but with a dark background of this world brought to light by our experiments. But this means that, in a way, this objective world is a product of our active intervention, and improved technique of observation. Here, too, then, we are brought face to face with the limitations of human understanding which we cannot overcome.

Goethe's struggle against the physical theory of colour, then, will have to be continued today on an extended front. Helmholtz said of Goethe 'that his theory of colour must be regarded as an attempt to save the immediate truth of the 'sense-impression" from the attacks of science'. To-day, this task is more urgent than ever. The whole world is being transformed by enormous extensions of our scientific knowledge and by the wealth of its technical applications but like all wealth, this can be a blessing or a curse. Hence many warning voices have been raised during recent years counselling us to turn back. Already, they say, a great scattering of intellectual effort has resulted from our negation of the world of direct sense-impressions and the division of nature into different sectors. Further withdrawal from 'living' nature will, so to speak, drive us into a vacuum where life will no longer be possible. When we are not advised simply to throw over all science, pure and applied, we are exhorted to develop science in close connection with daily experience. We are told that it is not sufficient to understand the laws governing all processes of the objective world but that it is essential to visualize at any given moment all the consequences of these laws in our world of the senses. In his constant dealing with nature in his own experiments, the scientist should become so familiar with observed phenomena that laws would appear merely a useful summary of his experiences. Thus the danger of completely separating the two kinds of realities is to be avoided by making the world of experiments as direct and 'living' as surrounding nature. But it is obvious from the start that the interrelations of nature can only be understood by a man who is thoroughly familiar with the manifestations of nature in the field concerned. There has never been progress and discovery without detailed knowledge based on experimental results. But the dangers of modern science are not surmounted in this way. For our experiments are not nature itself, but a nature changed and transformed by our activity in the course of research. To effect a real change would undoubtedly entail a complete abandonment of the whole of modern technology and science, which is linked with it. Nobody is in a position to say whether such a break would mean happiness or disaster for mankind. But however we may feel about this, one thing is certain. Such a break is impossible. We have to reconcile ourselves to the fact that it is the destiny of our time to follow to the end of the road along which we have started.

At the beginning of our modern era navigation flourishing and the daring feats of the circumnavigators of the earth opened up the possibility of the conquest of distant lands and of the return with immense treasures to their homelands. There may have been some doubt as to whether the new wealth would weight the scales equally with happiness and distress. Perhaps there were warning voices then who advocated a return to the more peaceful and less pretentious conditions of life of a previous epoch. But at such times warning voices resound unheard. The attraction of foreign lands and treasures can only come to its natural conclusion when these countries have been explored and their treasures have been distributed. Only then shall we have the vision to see more closely defined tasks, though they may be more important, and it is thus that science and technology will continue to develop in our time. Just as frontiers could not prevent the attraction of foreign countries, so no external obstacles will be able to prevent the progress of technology. Only nature herself can call a halt to our endeavours by showing us that the field to be conquered is not infinite. It is perhaps the most important trend of modern physics that it shows us the limits of our active attitude to nature.

Atomic physics took as a starting point the apparently natural supposition that our knowledge of the atom will, with increasing accuracy of observation, perfect itself more and more. Though atoms represented the final indivisible 'brick' of matter, they nevertheless appeared to be miniature parts of ordinary matter. The atom then, at least in our imagination, was endowed with all the macroscopic properties of matter. Only in the course of time was it recognized that the smallest particles, for instance electrons, could not themselves possess the 'sense-properties' of matter if they were to explain these properties on a larger scale. Otherwise the question of the reason for those properties would not have been solved but only moved one step further away. For example, if we say that a stronger movement of the atoms within differentiates a hot from a cold body, then an individual atom can be neither hot nor cold. Thus the atom was progressively divested of all its 'sense-properties'. The only properties which appeared for a long time to be retained were geometrical ones— the atom took up space and position, and had a definite movement. The development of modern atomic physics, however, has removed even these properties by showing that the degree to which such geometrical concepts can be applied to the smallest particles depends directly on the experiment in which they are involved. True, with a comparatively moderate demand for accuracy, we can speak of the position and velocity of an electron: true also that, compared with our daily experience, this accuracy is quite considerable. But measured by an atomic scale it is insufficient, and a law characteristic for this miniature world prevents us from determining position and velocity with the desired accuracy. Experiments can be done enabling us to determine, say, the position of a particle with great accuracy, but in the course of this measurement the particle has to be exposed to strong external influences which are responsible for a considerable uncertainty as to its velocity. Nature thus escapes accurate determination, in terms of our commonsense ideas, by an unavoidable disturbance which is part of every observation. It was originally the aim of all science to describe nature as far as possible as it is, i.e. without our interference and our observation. We now realize that this is an unattainable goal. In atomic physics it is impossible to neglect the changes produced on the observed object by observation. We decide, by our selection of the type of observation employed, which aspects of nature are to be determined and which are to be blurred in the course of the observation. This is the property which separates the smallest particles of matter from the range of our commonsense concepts. The supposition that electrons, protons and neutrons, according to modern physics the basic particles of matter, are really the final, indivisible particles of matter, is only justified by this fact. It would no longer make sense to visualize a three dimensional structure of these particles.

From what has been said we can conclude, along two different lines of thought, that the range of science and technology as we know it, is finite. On the one hand, our arrival, in atomic physics, at the final indivisible particles of matter should, in the not too distant future, lead to a complete survey of all the forces of nature yet to be exploited and hence of all possible technical possibilities. On the other hand, the way in which atomic phenomena are divorced from those of our everyday experience serves as an important example that in science the way in which a question is put and the method of research employed already singles out a finite and limited field from the abundance of physical phenomena. Previously, it appeared to be the task of science to describe the motion of bodies in space and to understand their regularity. Now we recognize that the range of atomic phenomena cannot be tackled in this way. When we ask of nature position and motion within an atomic system we destroy, through the impact of essential experimental measures, certain interconnections characteristic for a world of atomic size.

It is tempting to generalize these ideas and to recall Goethe's criticism of Newtonian physics. Goethe said that what the physicist observes with his apparatus is no longer nature. He probably meant to imply that there are further and more 'living' aspects of nature which are not accessible to this particular method of science. We are, of course, ready to believe that science, where it turns from inanimate to living matter, will have to be more and more careful in its interference in the course of an experiment. As our desire for knowledge also reaches out to higher, spiritual aspects of life, so we shall have to be content with a passive, contemplative kind of investigation. From this point of view, the division of nature into a subjective and an objective sector would appear an over-simplification of reality. It would be more to the point to imagine a division into many overlapping sectors, divided by the type of question we ask of nature and by the amount of interference which we allow during observation. In attempting such a classification in simple terms we are reminded of the classification of 'related aspects' as it appeared in the appendixes to Goethe's theory of colour. Goethe stressed that all the effects which we observe by experience are connected and continuous, yet the separation of one from the other is unavoidable. He classified them from low to high; accidental, mechanical, physical, chemical, organic, psychic, ethical, religious and of genius ('genial'). Seen in the light of modern science we might perhaps change some of the first delineations. For *mechanical* we might substitute all those phenomena accessible to classical physics, where a strictly causal space-time description can be given. The sphere of *chemistry* would include atomic processes and its scientific structure would be made plain by modern atomic physics. Besides these two categories we should not need a specific category *physics* as, in a sense, the previous two would be part of it. Neither should we allot a special category to *accidental* as accident plays a role precisely prescribed by natural laws. Thus the four lowest categories of Goethe's arrangement can be plainly understood in their scientific structure, their interrelations and their respective delineations. As far as the next category *organic* is concerned, modern biology believes it can recognize its limitations though indistinctly, and understand its inner structure. It is unlikely that anybody would, at this time, dare to define the remaining higher categories.

Dividing reality in this way into different aspects immediately resolves the contradictions between Goethe's and Newton's theories of colour. In the great structure of science, the two theories take up different positions. It is certain that an acceptance of modern physics cannot prevent the scientist from following Goethe's way of contemplating nature too. It would of course be premature to hope, on this basis, for an early return to a more direct and unified attitude to nature. It appears to be the task of our time to grasp, by experiment, the 'lower reaches' of nature and, through technology, to make them our own. In our advance in the field of exact science we shall, for the time being, have to forgo in many instances a more direct contact with nature such as appeared to Goethe the precondition for any deeper understanding of it. We accept this because we can, in compensation, obtain an understanding of a wide range of interrelations, seen with complete mathematical clarity. This must, undoubtedly, also be the basis and precondition for a proper understanding of the 'higher reaches'. Those who regard this as too great a sacrifice will, for the time being, be unable to devote themselves to science. They will only grasp that sense of science where, at the outer limits of present-day methods of research, science discovers its relations to life itself.

Perhaps we can liken the scientist who leaves the field of direct sense-impression in order to see nature as a whole, to a climber who wants to master the highest peak of a mighty mountain in order to survey the country below him in all its variety. The climber too must leave fertile inhabited valleys. As he ascends, so more and more of the country unfolds below him, but also life around him becomes more and more sparse. Eventually he reaches a dazzling, clear region of ice and snow in which all life has died and where he can only breathe with great difficulty, and only by traversing this region can he reach the top. But once he has reached it, in the few moments in which the whole country below him is visible with absolute clarity, he may not be so distant from life. We can appreciate that previous eras felt those lifeless regions to be only frightening wastes and an intrusion seemed an injury to some higher power which was bound to take bitter revenge against those who dared to approach them. Goethe, too, sensed an injury in the advance of science. But we may be certain that that final and purest clarity, which is the aim of science, was entirely familiar to Goethe the poet.

6

On the Unity of the Scientific Outlook on Nature[1]

We are witnessing a change in the external features of the world. The struggle for its reshaping is carried on with all our resources and absorbs all our powers. In such times, changes in the world of the mind, of which science is a part, automatically recede into the background. Yet the slow changes of human thought and desire have no less an impact on the external features of the world than great single events. A fundamental and lasting change, which has gradually matured in some fields of intellectual activity, can also be of importance on a world scale, in the shaping of our future. We may thus be justified, for once, in looking at our epoch from an unaccustomed angle. In the realm of science our times can be described as momentous. It appears that the various branches of science are beginning to fuse into one great entity and it is this unity I wish to discuss. The way in which this topic has been raised already implies the admission that up to now things have not been going too well.

I. Let us first turn to the initial stages of science at the beginning of the modern era. In the days when Galileo discovered the law of falling bodies and when Kepler studied the motions of the planets, there existed a single unified idea of nature, but it was not yet a scientific one. The picture of the world was still entirely determined by belief in a supernatural revelation as laid down in the Holy Scriptures. The scientist thought it was his

[1] Lecture delivered on November 26th, 1941 at the University of Leipzig.

task to recognize God's work in nature and to glorify it by understanding its harmonies in a scientific way. It is unlikely that either Copernicus or Galileo ever considered the possibility that the consequences of their scientific discoveries might lead to fundamental conflicts with existing religious views. That applied also to those theories which broke with traditional views and which caused conflict with the church. Even Kepler thought the study of the harmonies of the spheres, to which belongs his famous third law, was no more than a pursuit of the progress of God's creation. Thus we find at the end of the fifth book on Cosmic Harmony:

'I have endeavoured to gain for human reason, aided by geometrical calculation, an insight into His way of creation; may the creator of the heavens themselves, the father of all reason, to whom our mortal senses owe their existence, may He who is himself immortal . . . keep me in His grace and guard me from reporting anything about His work which cannot be justified before His magnificence or which may misguide our powers of reason, and may He cause us to aspire to the perfection of His works of creation by the dedication of our lives. . . .' This affirmation of faith certainly represented the basic attitude of early science and it was the church, in its struggle against the new theories, which instinctively felt the potential future dangers of this new science.

Only a few decades later, however, the scientist's task and hence also his concept of nature, had changed fundamentally. Attempts to bring mathematical order into the wealth of observations of nature, to clarify and explain them, had been more than successful, but, at the same time, the difficulties and the immensity of the tasks became more and more appreciated. The scientist at the beginning of the eighteenth century was no longer, like Kepler, just about to reach his goal of understanding God's plan of creation and of paying homage before the shrine thus unveiled. He stood on the threshold of an infinite stretch of virgin soil without visible end. This change of attitude cannot be better expressed than in the well known saying of the English scientist Newton:

'I don't know what I may seem to the world. But as to myself I seem to have been only like a boy playing on the seashore and diverting myself in now and then finding a smoother pebble or prettier shell than ordinary, whilst the great ocean of truth lay all undiscovered before me.'

The intervening period had seen the triumph of a new cognition; the new method had opened up to science a limitless expanse: simple processes of nature were to be unravelled by means of suitable experiments and the laws thus discovered laid down in mathematical language. This method could be applied to individual problems posed by nature and hence it was no longer a question of understanding a single interconnected whole but of a detailed analysis of many small specific connections. Of course Newtonian mechanics included a considerable part of nature accessible at that time to physical experiment. But there was also optics in which mechanical concepts could not be directly applied, and there was no question at all of investigations into animate nature. Mechanics could be a model and a basis for all other fields of science but the task of really 'ordering' all science according to it appeared to be endless.

The succeeding centuries have courageously tackled this task. The eighteenth century achieved decisive advances in the understanding of electrical phenomena. It laid the foundations of modern chemistry and achieved a series of important astronomical advances. It collected and systematized many experiences in the realm of animal and plant life. The nineteenth century raised our knowledge of Heat and of electric and magnetic phenomena to the level of Newtonian mechanics. In general, research broadened and deepened in most other fields of science to an extent far beyond previous achievements.

Such a development led inevitably to an atomization of science, where each section offered such a wealth of problems that no single individual could hope to master completely even a sub-section. It led also to the much complained-of specialization and a new evaluation of scientific work. Previously, the driving force of research had been the desire to understand the world in all its ramifications but as a whole—to re-live the Plan of Creation. To-day the scientist's pride is love of detail, the discovery and systematizing of the smallest revelations of nature within a narrowly circumscribed field. This is naturally accompanied by a higher esteem for the craftsman in a special subject, the 'virtuoso', at the expense of an appreciation of the value of interrelations on a larger scale. During this period one can hardly speak of a unified scientific view of nature, at least not as far as content is concerned. The world of the individual scientist is that narrow section of nature to which he devotes his life's work.

There is, it is true, a certain common scientific method and—as an expression of that method—a conception of the final aim of science, for which the model at least for exact science, was Newtonian mechanics. One had, from certain given data, to calculate the movement of nature, and many scientists were convinced that this task could be solved, at least in principle, in all fields of science. The most concise expression of this view of science at the time of rationalism was Laplace's fiction of the demon. He would be in possession of the complete data on the present state of the world and from this knowledge he could derive its whole future development.

The aim, then, was the creation of an edifice including all laws of nature which would make such a calculation possible, at least in principle. Whether all natural phenomena could, in the last resort, be traced back to the laws of mechanics or whether there might be yet other types of systems of concepts was left open.

The example of Newtonian methodology did, however, unify only the so-called 'exact' sciences. Quite different ideas were current among scientists concerned with animate nature. Vitalism, so prevalent in the second half of the eighteenth century, adhered to laws independent of physical and chemical interrelations. This was true in spite of the occasional references to electrical processes in connection with the 'elan vital', the force which distinguished animate from inanimate matter. It was in fact an *a priori* assumption that the laws of the processes of living matter were of a different character from the laws of

UNITY OF THE SCIENTIFIC OUTLOOK ON NATURE

physics. Most important of all—there never was a question of a mathematical formulation of these laws or the prediction of animate processes. In chemistry, too, it was at first widely believed that substances formed by living organisms were of a fundamentally different structure from those which originated in inanimate nature and which the chemist can synthesize from the elements in a retort. The possibility of relations other than those of 'exact' science were especially stressed and generalized by romantic natural philosophy. Important scientists have attempted, in vain, to introduce the type of law postulated for animate processes into processes of inanimate nature, e.g. astronomy. The romantics defended themselves against all attempts at explaining natural processes in terms of 'mechanical manipulation'[1] ('Stossen und Schlagen'). But these endeavours of the romantics could not prevail against the methodical assurance and transparent clarity of 'exact' science.

During the second half of the nineteenth century one could perhaps speak of at least a methodological unity of science. Wöhler's discovery had introduced the synthesis of organic substances from inorganic matter and this convinced the chemists that chemical reactions in living organisms were governed by the same laws as those in inorganic matter. From then onwards chemistry followed methodically the example of Newtonian mechanics and the success of the 'atom-hypothesis' made its contribution in spreading the ideal of a science based on the mechanics of elementary particles. In biology, vitalist views had been attacked by Darwin's theory of evolution and increasing attention was paid to analysing cause and effect. Even in medicine great successes had been achieved by an attitude of mind which likened processes in organisms to processes in a complicated machine.

In a way there existed thus a unified scientific view. Nature consisted of matter subjected, in conformity with natural laws, to change in time and space by action and reaction. Such changes took place by movement in space, or perhaps the internal movement of individual parts or again by a change of

[1] This is a reference to Oken as quoted on page 36. [F.C.H.]

material qualities (colour, temperature, tensile strength) which also depended on movement of the smallest particles, the atoms. We can regard such a view as an idealization of nature in which time and space are treated as independent categories into which events are projected as objective happenings. It is precisely this idealization on which Newton's mechanics is based and, as we have seen, mechanics was the methodological example for all science.

Although this view of nature had decisively advanced the development of science it was soon seen that it was incapable of creating a durable unity of its different branches. For the idealization just described hardly suited the concepts and problems of all the individual sciences. The system of chemical concepts had developed from an observation of material qualities and had become to a great extent independent of mechanical explanations. In biology scientists had to deal with processes of an altogether different kind which could be made intelligible by concepts such as growth, metabolism, heredity, etc. Finally no suitable place could be found in this view of nature for that great realm of reality comprising mental processes, and this was probably partly responsible for the much regretted division of mental activity into the sphere of science and the realms of art and religion. We can understand that this view of nature could never be fully convincing; nor could it prevent the disintegration of science into highly developed individual disciplines. It necessarily favoured a development in which the application of scientific thought to practical ends took the place of the 'universitas literarum'.

Though it cannot be said that this development has exhausted itself, there are some clear indications that the sciences are beginning to be drawn together more closely by new and different perspectives and there can be little doubt that the one-sided scientific view of the late nineteenth century is being replaced by new forms of thought.

II. The new process of the unification of science had its basis, however, not in the method but in the content of the individual branches. During the second half of the last century we had already in two instances achieved the fusion of different branches of science. Experience had shown that in increasing the heat of a body its smallest particles are always set into more rapid motion. Thus Heat and Mechanics became so closely linked that one could regard their phenomena as different manifestations of the same physical reality. Maxwell's famous theory traced back the theory of light to electro-magnetic processes. Light was shown to be an electro-magnetic wave process and Optics thus lost its status as an independent branch of physics. It had become a branch of Electricity and eventually part of general technology. Two individual disciplines had passed through the stages which are probably common to all branches of science. These stages could be sketched in the following way: for some considerable time fundamental relations within a certain field have to be classified by experiment and theoretical analysis. In so far as an understanding of these relations has been achieved—although it may be incomplete—technology can take over, and a considerable proportion of research is then devoted to practical applications. Eventually the 'complete text' of the laws of nature governing the field will have been found and the remaining scientific work will be solely directed towards the application of the laws to practical problems. In this way Maxwell's theory resolved all fundamental problems of optics, and scientific interest was concentrated on technical questions such as the manufacture of optical instruments.

A consistent development of this idea has, ever since the beginning of the twentieth century, moved atomic physics into the focus of scientific interest. This discipline had—since its origin in early classical antiquity—assumed the exalted task of relating the behaviour and all the properties of matter to the movement of its smallest particles, the atoms. It had attempted to derive all physical and chemical disciplines from a common root. Atomic physics posed the problem in this way: visible qualities of matter such as the taking up of space, strength of materials, colour, chemical properties, etc., are attributes of macroscopic matter (*Materie im Grossen*) but they cannot be attributed in the same way to the smallest indivisible 'bricks' of matter. Otherwise we could not perceive one and the same matter existing in different forms (e.g. water as ice, water and steam). These macroscopic qualities are produced only by the movement and the mutually acting forces of the smallest particles.

In the nineteenth century the development of chemistry supplied a firm foundation—an 'atom-hypothesis'. We know now that a piece of a chemical element, of carbon for instance, can be divided into ever smaller parts until we finally arrive at the smallest unit characteristic of this element. This is the atom of the element—in our case the 'carbon atom'. In a chemical compound different types of atoms, i.e. atoms of different elements are ranged together into an atom group, the 'molecule'. Hence the molecule is the smallest unit of a chemical compound. This conception helps us to account roughly for the chemical properties of matter.

But from another point of view the task of atomic physics appeared insoluble. Chemistry dealt only with specific qualities which always recur in the same way in the same kind of matter. They are properties which show a special stability in face of all kinds of disturbances. A piece of gold always shows the same reddish colour irrespective of how it was obtained and how it had received its shape. Such a stability of external properties is, however, alien to mechanical systems. The orbit of a planet round the sun would be permanently affected by an external disturbance, say the intersecting of this orbit by a comet of large mass. After the gradual fading of the disturbance, the planetary system would not return to its original configuration. It would require very strange suppositions concerning the properties of atoms to explain, mechanically, such a type of stability.

In the course of time, this difficulty proved to be the real central problem of atomic physics and its solution only became possible on the basis of the quantum hypothesis as enunciated by Planck in the year 1900. We can only touch upon the historical development of quantum theory and Bohr's conception of atomic structure. In his investigations on the radiation of hot bodies Planck had first discovered a strange discontinuity of the energy content of the atoms. It appeared as though a small radiating system could have only quite definite, discrete energy values. Later Rutherford developed from his experimental work the idea that an atom can be likened to a small planetary system in whose centre is a positively charged atomic nucleus, embodying practically the entire mass of the atom. Round this nucleus circle negative electrons. The stability of this planetary system could be explained by Bohr a few years later by means of Planck's quantum hypothesis, and finally, a quarter of a century after Planck's discovery, the exact mathematical form of the laws governing atomic structure were found.

Quantum theory did in fact satisfy all the demands which, within the limits of our present knowledge, could be made on atomic physics. The theory enabled us, at least in principle, to calculate—and to that extent 'explain'—the properties of macroscopic matter. In the case of a few very simple substances, such as hydrogen, we have succeeded in calculating with great accuracy the most important chemical properties, the colour in discharge tubes, phenomena at low temperatures and other related properties. These calculations have even brought to light some phenomena which had been overlooked by the careful work of the experimental physicist. In the case of many other substances, quantum theory can supply at least a qualitative explanation of their properties such as, for example, electrical conductivity of metals or the structure of crystals. Thus we are perhaps justified in believing that we have reached a level of research comparable to that of the knowledge of the mechanics of the heavens after Newton. We may say that we are capable of a quantitative 'calculation' of the properties of matter in all cases where mathematical complications do not prevent the execution of this task in practice.

A heavy price had, however, to be paid for the achievement of this ambition. It meant, in the simplest form, the loss of just that nineteenth century scientific conception of nature or, expressed more accurately, the loss of that conception of reality on which Newton's mechanics rested.

This was because quantum theory made the atom into something inaccessible to our senses or our imagination, unlike objects within our daily experience. An atom or, more correctly, the smallest unit of modern nuclear physics, an electron no longer displays 'in itself' ('*an sich*') even the simplest geometrical and mechanical properties but it shows them only to the extent to which they can be made accessible to observation by external interference. Different observed properties of an atom are complementary in the sense that the knowledge of one particular property can exclude the simultaneous knowledge of another property. This strange kind of reality of the atom or the electron carries with it various important consequences. The behaviour of an atom in many experiments can be described by means of mechanical concepts: we can, for example, speak of the track of certain particles. In such experiments the laws of classical mechanics always provide a correct account of the event concerned. Hence, we can say that the laws of classical mechanics apply to all those atomic processes in which they can be directly checked. There are also, an the other hand, experiments in which it is necessary to use non-mechanistic concepts for a description of the state of an atom, e.g. concepts which express the chemical properties of an atom. In such cases no use can be made of mechanistic descriptions and the question as to whether the laws of mechanics 'apply' is irrelevant. Mechanical and chemical properties are mutually exclusive. This is clearly expressed in the mathematical formulation of the quantum laws and makes possible the peculiar non-mechanical stability of atomic systems on which is based our knowledge of macroscopic matter (*Materie im Grossen.*).

These facts demonstrate the finality and assurance of classical theory which apparently cannot be shaken by any new experience, and which holds good wherever its concepts apply. On the other hand, nature makes provision for relations of quite a different kind by forcing us to create some external disturbance in the course of each observation and thus withdrawing from our grasp an apprehensible picture of the atom. An atom can no longer, without reservation, be 'objectively' described as an object in space changing in time in a definable manner. Only the results of individual observations can be objectively described but they never present a complete and apprehensible picture. It follows that the conception of reality on which Newton's mechanics was based was too narrow and had to be replaced by something broader.

Previously, physics had attempted to treat processes accessible to our senses as secondary and derived and to explain them in terms of events on an atomic scale (*in kleinen*).

These events were considered to be the 'hidden' objective reality. However, we now recognize that events accessible to our senses (with or without the aid of scientific apparatus) can be considered to be 'objective'. That is to say, we can justifiably claim that an event observed by us has 'objectively' taken place. But atomic processes cannot always be represented as objective events in time and space. Only a reversal, if I may express it in this way, of the order of reality as we have customarily accepted it, has now made possible the linking of chemical and the mechanical systems of concepts without contradiction.

Atomic theory has thus joined physics and chemistry into one great and unified science. We may ask ourselves what has been the practical effect of this new unity on the individual parts of science and what influence has it already exercised on our scientific conception of nature?

We might have thought that the new situation would necessarily lead to an extraordinary upsurge of chemistry, since all fundamental problems of chemistry, e.g. the nature of chemical forces, had now been solved by atomic physics. On closer inspection it soon becomes apparent, however, that chemistry had long since left the realm of research into fundamental relations in favour of that of their practical application. The problem of the nature of chemical forces, once the central problem of chemistry, has been so completely relegated to the background that no chemist needs to take any notice of it when dealing with some small, though in practice perhaps important, question. A fundamental solution could be of little use in a particular question since real theoretical treatment based on atomic theory would meet, in most cases, with insurmountable mathematical obstacles. Therefore the influence of modern atomic physics has so far been limited to only a few sectors of chemistry, and the value of the new concepts is only gradually assuming importance in more general problems such as the theory of valency. Atomic physics has been much more fruitful in astro-physics, i.e. the theory of the physical structure of the stars. Many questions concerning the atmosphere of stars and the origin of internal energy could only be tackled on the basis of atomic physics.

If, finally, one looks for the effects of atomic physics on the basic problems of physical reality we have to admit that the concept of reality, achieved in the unity of 'exact' science, has not everywhere been accepted without resistance. This resistance came in some cases from particular sciences which were not inclined to sacrifice traditional and proven concepts for the sake of a higher unity. It also came from certain theories of perception which could not accommodate the new situation with which nature confronted us at an atomic level.

But there is little reason to be anxious about the fate of this newly-won unity. Resistance is, after all, directed not against the results but only against their interpretation, and thus does not touch the core of content of the new concepts. In a similar way violent battles took place in world history when the time came for a unification of various small states, but in the end the struggles led to unity. The past is always of such importance for us humans that we want to take over into a new unity as many as possible of the old and familiar values. So in science it would have been too much to expect that a fundamental change such as the unification of physics and chemistry into a single scientific system could have been accomplished without the adoption of new and unfamiliar forms of concepts.

Surely it would be better, in spite of the unaccustomed difficulties involved, to come to terms with the new forms of thought and to see what fields might thus be opened to us.

III. Let us turn first of all to the science which ranks immediately above physics and chemistry, the science of life. For a long time there have been two fundamentally opposed trends in biology. They can be conveniently distinguished as vitalist and mechanist. Vitalism postulates that relations characteristic for living processes are basically and qualitatively of a different kind from physical and chemical laws. Organisms can be described in terms of growth, metabolism, propagation, adaptation, tendency to healing, etc., concepts which do not occur in physics or chemistry. It could, of course, be thought that properties giving rise to these concepts arise from physical and chemical laws. But our knowledge of these laws provides no grounds for such a supposition and we have to admit that a living organism would be a very improbable structure from the standpoint of physical and chemical laws. It may be compared with crystals, which are equally improbable when seen from the point of view of classical physics as applied to electrons and atoms.

The idea that life has its own laws and nomenclature is opposed by the so-called 'mechanistic' school. They say that organic processes which can be investigated always appear to conform to the known laws of physics and chemistry, and that deviations from these laws have never been observed in living matter. Secondly, there is no room for relations of any other kind since physics and chemistry completely determine all properties of matter. In fact, quite extraordinary progress has been made in the past hundred years in the physical-chemical explanation of organic processes. I may mention the question of heat exchange (*Wärmehaushalt*) in organisms, electrical effects in the nervous system, the chemistry of hormones, etc. All previous experience seems indeed to bear out that the known physical and chemical laws hold without exception in the realm of living matter. On the other hand, these laws leave no gap which could be completed by relations of another kind, and this state of affairs is in no way altered by quantum theory in which statistical laws play so great a part. There remains thus—at least apparently—the much discussed contradiction between individual processes in organisms, which can be entirely explained by physics and chemistry, and the characteristic processes of life 'as a whole'. The common saying that 'the whole is more than the sum of its individual parts' expresses this contradiction but it does not resolve it.

The problem we have just posed appears in an entirely new light if we make use of the methods of thought of quantum theory and if we, like Bohr, take the theoretical situation of atomic physics as an example of method. In quantum theory too there apparently existed, at first, a contradiction between classical physics on the one hand and chemical concepts on the other. The first completely determined the properties of a system from its initial conditions and applied wherever these could be checked, the second led to a system of concepts which had no immediate connection with classical physics. The contradiction was resolved by our knowledge that a situation which could be described in chemical terms excluded the accurate knowledge of the mechanically determining conditions. This exclusion arises automatically through the disturbance which, according to natural laws, is inevitably implied in every observation. We can imagine a similar situation in biology. The statement that 'a cell is alive' could include an accurate and complete knowledge of the conditions which determine its physical structure. The achievement of such a complete knowledge would probably necessitate such drastic interference (e.g. the use of X-rays) that the cell under observation would be destroyed. At least the methodological example of quantum theory can demonstrate that there is no necessary logical contradiction between the basic thesis that 'the physical-chemical laws apply without qualification in living nature' and the vitalist thesis that life has its 'own' laws.

This does not yet, of course, solve the problem, and research has for some decades explored the borderlines where a solution is likely to be found. The properties of an organism as a whole are of little help since their physical and chemical relations cannot be grasped in all their complexity. When they can be understood, their physical and chemical properties are obvious. The real problem of living organisms as a whole lies in the very reason for the origin of such complex formations and this question immediately leads us to those of growth, cell division, the doubling of chromosomes and genes, i.e. the borderline between biology and the chemistry of large molecules.

In this field, the results of the new atomic physics can be used not only for their method but also for their content, and the special aspects of quantum theory which relate to the theory of perception gain in importance not only in method but also in content. Genetical investigations into the frequency of mutations, for instance, seem to indicate that under certain conditions an event on an atomic level, such as the release of a single chemical link in a chromosome of a cell nucleus, can cause changes in the whole future development of the organism. In such cases the statistical laws of quantum theory assume a direct practical importance for the behaviour of a living being

Studies on the borderline between the chemistry of albumens and the biology of the smallest elementary units will thus—apart from all considerations of principle—first of all have to exploit to the full the concepts of physics and chemistry, in order to establish just how far they can be used for a description of living processes. In doing this we are aware that the natural laws may themselves prove to be a barrier, and this would prevent us from neglecting those other aspects of life from which vitalism once drew its strength and which impress upon the observer's mind what has been called 'reverence before Life'. The change in the order of reality which has taken place within quantum theory has also brought the biological branches of research, whose subjects are those other characteristic aspects of the process of life, much nearer to the 'exact' sciences. It means that, apart from specific borderline research, certain common thought relations have been established between two previously entirely separated fields of science.

The developments of the last decades have thus drawn biology, physics and chemistry more closely together. A real fusion of the three subjects into a unity of content could, however, only be achieved by fundamental extensions of our knowledge of the processes of life. But there seems to be already a beginning of a methodological unity which is no longer supported by the desire to explain all processes by the example of Newtonian mechanics. The methods of thought which have arisen from atomic physics are wide enough to offer scope for all the various aspects of the problems of 'life' and the scientific endeavours connected with them.

Of course such a methodological unity cannot justly be called a unity of the scientific conception of nature. Such a conception must, at least in principle, be able to accommodate *all* parts of nature and it must be able to allot a definite place to each sector of reality. It was precisely this demand which so clearly demonstrated the shortcomings of the views based on classical physics. In that picture of nature the mental world figures, so to speak, only as the opposite pole of a material reality incapable of accommodating it within its bounds. The structure of the concepts of classical physics was too rigid to assimilate experiences of a new and different type without forcible interference.

The order of reality of modern atomic physics is perhaps better able to allow for different systems of concepts. It does so because it makes phenomena 'objective' without the demand for a 'thing in itself' capable of description in terms of everyday experience. Other systems of concepts may be used for descriptions of other types of reality and it is to be hoped that one day they will be united, clearly and in detail, with those already better known.

Thus, the changes introduced by quantum theory have affected the position of theories of perception in such a way that those aspects of reality characterized by the words 'consciousness' and 'spirit' can be related in a new way to the scientific conception of our time. Classical physics was built on the firm foundation of the recognition of the objective reality of events in time and space which take place according to natural laws independent of mental activity. This means, of course, that they, in their turn, apply only to such 'objective' processes. Mental processes appear to be only an image of this objective reality which is separated from the world of time-space relations by an unbridgable gap. Improved modern technique of observation and the enrichment of positive knowledge resulting from it has finally forced us to revise the fundamentals of science and has convinced us that there can be no such firm foundation of all perception. After all, our idea of a world moving in time and space is only an idealization of reality dictated by our desire to see the world, as far as possible, objectively. Quantum theory uses a different idealization, less obvious and complying to nothing like the same extent with our desire to see things objectively but it enables us, in compensation, to understand completely the laws governing chemical changes. Chemical processes cannot be related to the physical behaviour of the smallest particles, within the framework of the conceptions of reality of classical physics, and we are thus prepared for other occasions when we again find that peculiar complementary relation between different aspects of reality.

Of course, we cannot assume such simple proportions as: biology relates to chemistry, as chemistry to physics. It would probably be more correct to say that a completely new level of perception and understanding has to be achieved in the transition from an aspect of reality already 'understood' to one still new. Such a step may be as difficult as was the advance from classical physics to atomic theory.

Yet, having said this, we probably understand now, better than before, that there exist apart from the phenomena of life, still other aspects of reality, i.e. consciousness and, finally, mental processes. We cannot expect that there should be a direct link between our understanding of the movement of bodies in time and space, and of the processes of the mind, since we have learnt from science that our mental approach to reality takes place, at first, on separate levels which link up, so to speak, only behind the phenomena in an abstract space. We are now more conscious that there is no definite initial point of view from which radiate routes into all fields of the perceptible, but that all perception must, so to speak, be suspended over an unfathomable depth. When we talk about reality, we never start at the beginning and we use concepts which become more accurately defined only by their application. Even the most concise systems of concepts satisfying all demands of logical and mathematical precision can only be tentative efforts of finding our way in limited fields of reality.

Thus we are no longer in the happy position of Kepler, who saw the interrelations of the world as a whole as the will of its creator, and who believed himself, with his knowledge of the harmonies of the spheres, to be on the threshold of understanding the Plan of Creation. But the hope for a great interconnected whole which we can penetrate further and further remains the driving force of research for us too.

7
Fundamental Problems of Present-day Atomic Physics[1]

Practically all public discussion of atomic physics is in fact concerned with atomic technology, i.e. the application of the enormous energy of atoms to weapons of war or to machines. The real science however, of which this technology is but a branch development, is much less known to the general public. Occasionally there may be reports of the success of a British scientist in discovering a new elementary particle, or of new knowledge of the inner atomic forces gained in experiments with a new giant cyclotron in California, or again of Stalin Prizes awarded to two Russian scientists for their work in high altitude laboratories in the Caucasus. But the real aim, the common bond linking all the efforts of men of different nations and making them part of a pattern, this aim is hardly ever discussed. And yet this is precisely the object of atomic physics for the physicist. For him there is ever present in his work the centuries-old desire for a unified understanding of the world, and he judges every discovery, at least unconsciously, on its ability to bring him nearer to the goal of his ambition. That is why I should like to speak to you to-day about those fundamental ideas which combine various experiments and theories into atomic physics. I should like to explain what we are hoping for in our work and what will have happened when our hopes and wishes have been fulfilled.

[1] Lecture delivered at the Eidgenössische Technische Hochschule, Zürich, on July 9th, 1948.

PROBLEMS OF PRESENT-DAY ATOMIC PHYSICS

To get an understanding of the basis of atomic physics we shall have to follow, step by step, the ideas which, two and a half thousand years ago, had led Greek natural philosophy to atomic theory, and we shall then have to make an attempt at finding a connection with these fundamental ideas even in the advances of the most modern atomic physics. It will therefore be no digression if I first outline briefly the pre-history and the history of atomic theory.

At the beginning of Ionic natural philosophy we find the famous statement of Thales of Miletus that water is the origin of all things. This statement which appears so strange to us to-day, contains as Friedrich Nietzsche had already pointed out, three fundamental ideas of philosophy. First, the idea that there is an origin of all things, then that this question has to be answered rationally, and thirdly, that it must in the final resort be possible to 'understand' the world through a unified principle. These three implications are the more remarkable since it was, at that time, not at all an obvious step to look for the origin of things in something material rather than in life itself. Thales's statement is the first to contain the idea of a homogeneous fundamental substance of which the world is to consist, although the word 'substance' had certainly not the purely material sense which we, to-day, attribute so easily to it.

If there were only one such substance, then it would have to fill up all space uniformly and indiscriminately and the existing large variety of phenomena could never be explained. For this reason the philosophy of Anaximander, a pupil of Thales who also lived in Miletus, was based on a fundamental polarity, the contrast between Being and Becoming. A homogeneous existence ('*Sein*') gave rise to change, to Becoming. This, in turn, represented in a sense a corruption of the pure Being. It did so by shaping the play of the world by means of hatred and love. In Heraclitus's philosophy, Becoming assumed prime importance; fire became the basic element, that which moved, but it also represented Good and Light; war was the father of all things. Later, especially as a result of Anaxagoras's influence, the idea gained ground that the world consisted of several elementary substances which were believed to be homogeneous and indestructible; only their mixing and separation produced the variety of life. Empedocles adopted the famous elements earth, water, air and fire as the four 'basic roots' ('*Stammwurzeln*').

It was from this position that Leucippus and Democritus of Abdera effected the transition to materialism. The polarity of being and non-being was made worldly and became a contrast of Full and Empty. Pure Being contracted to a point, but it could repeat itself any number of times; it became indivisible and indestructible and hence it was called 'atom'. The world was reduced to atoms with empty space between them. A mixture of elements was thought to be like a mixture of two kinds of sand. The relative position and movement of atoms determined the qualities of substances and were thus responsible for variety in the world. Up to that time space had been conceived as something impossible *without* matter, but as something suspended *by* matter. Now, materialist philosophy endowed it with a certain independence, it became, as empty space between atoms, the carrier of geometry—that is, responsible for the whole wealth of shapes and all the varied phenomena of the world. The atoms themselves had no properties, neither colour, nor smell or taste. Properties of substances were produced indirectly by the relative position and movement of atoms. In Democritus we find these statements.

'Just as tragedy and comedy can be written using the same letters, so, many varied events in this world can be realized by the same atoms as long as they take up different positions and describe different movements.'[1]

'Sweet exists by convention, bitter by convention, colour by convention; atoms and Void (alone) exist in reality'[2]. . . .

Thus, atomic theory realized Thales's fundamental demand that nature must be capable of interpretation in terms of a unified principle by recognizing only one basic substance, one

---
[1] This is a summary of a passage in Aristotle 'De generatione et corruptione' A 1.314a21 ff. on Leucippus and Democritus.
[2] Ancilla to the Pre-Socratic Philosophers (translated by Kathleen Freeman) Blackwell, 1948) (68/9 p. 93).

fundamental form of Being, namely the atoms. This pure existence was contrasted with form and movement which personified the process of Becoming and caused the totality of events in nature. Plato, who in *Timaeus* accepted the ideas of atomic theory, distinguished five kinds of atoms, which differed in shape, and which were supposed to correspond to five basic substances. This assumption of five types of atoms could at first sight appear as a retrograde step, but basically Plato thought only of a single entity which happened to appear in different shapes. Nature's variety was the result of the diversity of mathematical structures. The whole wealth of life was reflected in the wealth of geometrical shapes which, themselves, were formed by that which really existed, the atoms.

I have briefly outlined this historical development because it makes quite plain the fundamental aim of atomic theory. It is to explain that the world consists, in the last resort, of a homogeneous substance, and that it is based on one unified principle. The multiplicity of phenomena has, somehow, to be related to the multiplicity of mathematical structures. Later developments add to these ideas the important conception of unalterable natural laws governing all events. Thus mathematical structures have been made to reach into the future, and allow us to predict future events. But these later developments adopt, almost unchanged, the basic ideas of atomic theory, and even at the present day they retain their creative power.

Before dealing with our modern problems from the point of view of those basic ideas, I should like to pursue their historical development a little further, because only such a background will enable us to understand the sense of the endeavours of our own time. At the beginning of the modern era (*Neuzeit*) the conception of basic substances developed from chemical experiences. Thus, since the seventeenth century, substances which could not be chemically further sub-divided, ranked as basic elements and all matter was to consist of them. We now know about ninety-five chemical elements which make up about half a million chemical compounds found in nature. To every such element was attributed a type of atom such as the carbon or oxygen atom, which themselves were thought to be indivisible and indestructible. The compound was formed by the arrangement of atoms of different elements in atomic groups, the so-called molecules, such an atomic group then representing the smallest unit of the chemical compound concerned.

This atomic physical interpretation of chemistry was finally successful at the end of the eighteenth century and henceforth it formed the basis for the great advances of chemistry, yet we can say that this victory of atomic theory did not do full justice to its basic conception. The world was supposed to have consisted, in the last resort, of a unified substance. But this basic demand had been lost sight of, for the assumption of nearly 100 different elements of whose mixtures all matter was to have consisted, implied a degree of complication totally opposed to the basic aim of atomic physics. In spite of this, such great successes were achieved that an atomic interpretation of chemistry was generally accepted. After all, it was an indisputable fact that chemical elements could not be further subdivided or transformed by chemical means.

However, as early as 1815 the Englishman, Prout, attempted to break through such views when he defended the hypothesis that all elements consisted ultimately of hydrogen. He formed this thesis as a result of observing atomic weights, which could for the first time be determined with reasonable accuracy. The atomic weights of many of the lighter elements were, fairly accurately, integral multiples of the lightest one, i.e. hydrogen. For instance, a helium atom is almost exactly four times as heavy as a hydrogen atom, so it was tempting to believe that an atom of helium consisted of four atoms of hydrogen. But it was another hundred years before we could be certain that the atoms of chemistry were not the final indivisible units of matter or, in other words, they were not really what the Greeks meant when they used the word atom.

Faraday's investigations, his discovery of the electron (i.e. the atom of electricity and radio-active radiation) led us finally to Rutherford and Bohr's famous atomic model and thus introduced the latest epoch of atomic physics. For almost forty years we have known that, with certain reservations, an atom of a chemical element has to be imagined as a planetary system on a minute scale. The largest part of its mass is concentrated in the atomic nucleus which is positively charged and whose diameter amounts to about 10^{-5} of the diameter of the atom. Round this nucleus circle the much lighter electrons whose number is just sufficient to neutralize the charge of the nucleus. The diameter of the outermost planetary orbit amounts in most atoms to about 10^{-7} millimetres. The reservations which I mentioned earlier are concerned with the basic difficulty of describing processes in an atom in the imagery of every-day language. It is true we know the natural laws which govern the movement of electrons round the nucleus so well that we can formulate them mathematically with complete accuracy, but we can only very roughly translate these laws into an imagery that can be visualized. This is because Planck's quantum hypothesis, on which these laws are based, contains an aspect which is not apprehensible in principle.

The shells of all atoms consist then of the same 'substance', namely of electrons, the lightest, negatively charged elementary particles. The diversity of types of atoms is a result of the diversity of atomic nuclei, which cannot be affected by chemical means. But we can bombard them with other elementary particles of high velocity and we find, as has been expected for some time, that the nucleus itself is composite, and also that one atomic nucleus can be transformed into another atomic nucleus. For some fifteen years we have known that all nuclei consist of two kinds of elementary units which we call protons and neutrons, protons being identical with the lightest nuclei i.e. those of hydrogen, while neutrons are electrically neutral elementary particles of about the same mass as protons. We can tell how many protons and how many neutrons every atomic nucleus contains: thus, the nucleus of a hydrogen atom consists of one proton, the helium nucleus of two protons and two neutrons, the heavy nucleus of the uranium atom of 92 protons and 146 neutrons. The number of protons present determines the charge of the nucleus and hence the chemical properties of the atom.

PROBLEMS OF PRESENT-DAY ATOMIC PHYSICS

The discovery that all atomic nuclei consisted of the same units led immediately to a problem soluble at least in theory, i.e. the artificial composition and decomposition of nuclei. Since Hahn's famous discovery that neutrons can cause the disintegration of uranium nuclei, the artificial disintegration and building up of nuclei has become an important branch of modern technology; we are now really able to transform one chemical element into another.

If we compare the present state of atomic physics with that of 150 years ago, we can immediately say that our modern views are much closer to the fundamental aim of atomic theory, which was an explanation of nature based on one homogeneous substance. In place of the hundred odd independent basic chemical substances we now have only three, which should more accurately be called three fundamental forms of matter, whose atoms we name electrons, protons and neutrons. All matter, dead or living, consists of these three kinds of elementary particles and of nothing else. Qualitative differences are caused by different arrangements and relative positions of these three basic units. The multiplicity of possible phenomena finds its reflection in the multiplicity of mathematical structures which can be realized with these three basic forms of Being.

This last point is characteristic not only of atomic physics but of the whole of exact science and I should like to treat it in greater detail, using chemistry as an example. We know accurately the laws which govern the movement of electrons round the nucleus. Hence every possible state of atoms, say in a complex molecule, must correspond to a solution of the equations representing those natural laws. Thus our mathematical formulations are richer in content than those of the Greeks; we are no longer restricted to geometrical structures but we use complicated systems of differential equations which can, especially in the case of atomic physics, be defined in multidimensional space. The totality of solutions of such equations corresponds to the totality of all possible states of atoms, the wide variety of possible chemical compounds being depicted by the totality of possible solutions of Schrödinger's differential equations.

However, in considering three basic substances, i.e. three kinds of elementary particles—electrons, protons and neutrons—as the component parts of all matter, we have not altogether covered the programme of atomic physics. We are reaching here the real aim of modern atomic physics. If only these three elementary particles existed, we could rest satisfied in the belief that there are three fundamentally different sorts of matter which can no longer be transformed into one another or related to one another. But in reality there are yet other forms of manifestation of matter, the most important being radiation. Since the famous formula of relativity theory has linked energy and mass, we know that every form of energy also possesses mass and that it can therefore be called a form of matter. According to Planck and Einstein, energy in radiation is concentrated in so-called light-quanta which can also be regarded as some kind of elementary particle. But beyond that, still other elementary particles have been found. In the early thirties Anderson discovered the positive electron which can be created in the transformation of radiation into matter when, occasionally, a high energy light quantum, for instance in X-radiation, passes close to a nucleus and produces a negative and a positive electron. A little later, Anderson found a further elementary particle which was a result of cosmic radiation in the atmosphere. It is about 200 times heavier than an electron and now goes by the name 'Meson'. Mesons have, however, a very short life, disappearing after something like a millionth of a second and changing into one electron and one other neutral elementary particle. Finally there have recently been discovered still other elementary particles which also have very short lives.

In the face of this development of atomic physics in the last few years it may easily appear as though atomic theory is again losing sight of its fundamental aim, as though the assumption of three basic substances was again being replaced by more complicated assumptions. This question raises immediately the problematical nature of modern nuclear physics. Our present concepts still seem to be too simple: there are many indications that further elementary particles exist which have not yet been observed because they have an extremely short life. Also, another important fact has been found experimentally: elementary particles can change into one another, and the characteristic of indestructibility no longer applies in the old sense, e.g. a collision between a neutron and a proton can produce a meson. This is a process characteristic in general for the collision of two elementary particles of high energy. In such an impact some new elementary particles are frequently formed and this happens the more, the higher the total energy available is. The process is best described if we say that the total available energy of the impact is used in a statistical manner to form elementary particles, and that it is distributed among these particles. The particles thus created have a definite mass and other definite properties, some of them being well-known elementary particles. Particles of the same kind are always identical in their properties and to that extent they are uniform; but they can be transformed into one another.

This step which has only been accepted in the last few years does, however, take us close to the real aim of atomic theory. Just as the Greeks had hoped, so we have now found there is only one fundamental substance of which all reality consists. If we have to give this substance a name, we can only call it 'energy'. But this fundamental 'energy' is capable of existence in different forms. It always appears in definite quanta which we consider the smallest indivisible units of all matter and which, for purely historical reasons, we do not call atoms but elementary particles. Among the basic forms of energy there are three specially stable kinds: electrons, protons and neutrons. Matter, in the real sense, consists of these with the addition of energy of motion. Then there are particles which always travel with the velocity of light and which embody radiation, and finally other forms with a short life of which only a few have been discovered so far. The variety of natural phenomena is thus created by the diversity of the manifestations of energy, just as the Greek natural philosophers had anticipated. If we wish to understand all these manifestations we should be able to represent them in mathematical form, in the last resort, simply by the totality of solutions of a system of equations, and it is here that we come up against the decisive problem of modern atomic theory. It is that the mathematical formulations which describe the properties of elementary particles are not yet entirely known, but only knowledge of them will enable us to predict the results of experiments, that is, to master events in the same manner as physics has done up to now. We can also see that little has been gained by the definition of one fundamental substance since all the wealth of phenomena is harboured by its manifestations. What understanding of matter we have achieved has finally been written down in mathematical equations, for no other language can dispose of such a wealth of expressions. Thus we can say that the real task of atomic physics in the next few years or decades will remain the experimental discovery and mathematical formulation of those natural laws which determine all the properties of elementary particles and their combinations. The discovery of a new particle in cosmic radiation will, for instance, provide us with new information about those laws. If extensive mathematical investigations are carried out to study the properties of bi-linear forms (used to represent observable quantities in modern nuclear theory) then we may discover something of the mathematical formulations which, in future theory, will describe also the properties of elementary particles.

Perhaps I may say a few words here about the peculiar difficulties with which we have to contend. In any mathematical description of nature we have to introduce certain mathematical symbols which are used for formulating equations which, in their turn, represent natural laws, e.g. we use symbols for the position and velocity of particles in Newtonian mechanics. If, however, we make use of any of the common symbols, such as the co-ordinates of a particle, we are already tacitly implying the existence of a given particle. Yet it is the decisive point of this last stage of atomic physics that particles can no longer be taken for granted, since we want to understand their existence and their properties; thus we cannot sensibly assume co-ordinates and mass of definite particles. The question arises as to what we can use. We have not yet really developed the mathematical tools which would grasp the complex events on a nuclear scale. It could, of course, be said that though particles cannot be definitely *given* but have to be *determined* they will nevertheless possess position and mass so that these variables can in any case be introduced into the equation. But is it really true that particles have position? They certainly have position with a considerable degree of accuracy, but are there not likely to occur similar and perhaps even more stringent limitations of accuracy such as have appeared in quantum mechanics? We can see what great difficulties atomic theory has to master here. Yet it is quite conceivable that in the not too distant future we shall be able to write down a single equation from which will follow the properties of matter in general.

If we really succeed in this, atomic theory will have reached its ultimate goal, and it may be worth finding out just what we shall have achieved. First, we shall have understood the unity of all matter in the sense in which the Greeks used that phrase. All matter consists of the same substance, energy which manifests itself in different forms, and the totality of these forms is governed by the totality of solutions of a system of equations. That would mean that the results of experiments in atomic physics could be predicted, at least in principle. We can also assume that these mathematical forms would not only apply to the branch of atomic physics, since even present-day atomic physics contains, at least in principle, chemistry, mechanics, optics, heat and electricity, and this will certainly apply to the atomic theory of the future. When we use so frequently the expression 'in principle' as a limitation, we mean that in most cases the complete mathematical mastery of a problem is technically impossible, for our mathematics just cannot cope with such complications. It is therefore not at all certain that the solution of the fundamental problem will yield much that is of use in practical application. But the expression 'in principle' means also that a solution of the fundamental questions can be of use in all cases where we have to deal with a solution of a specific problem.

There are, however, two counts on which we should question

how far modern atomic theory would have satisfied Greek philosophers. The mathematical forms in the minds of the Greeks were geometrical shapes which could be visualized and which were, so to speak, traced in empty space by the atoms. Can the mathematical forms of our atomic theory be similarly visualized? Secondly, Greek atomic theory set out to describe the properties of all reality, mental processes and living organisms as well as purely material processes. Democritus said: 'there are only atoms and empty space.' Does modern atomic theory relate only to a narrower field and are we to assume that there exists, apart from atoms, something else—for instance, a soul? Or does our theory also maintain that 'there are only atoms and empty space'?

The first question has often been dealt with. In fact our modern atomic physics is much less apprehensible than earlier scientists had hoped. But we have been reconciled to this because nature has taught us that it is closely and consistently linked with the existence of atoms. We could put it like this, though it would be a little inaccurate: anything that can be imagined and visualized cannot be indivisible. The indivisibility and homogeneity, in principle, of elementary particles makes it quite understandable that the mathematical forms of atomic theory can hardly be visualized. It would even seem unnatural if atoms lacked all the general qualities of matter like colour, smell, taste, tensile strength, and had yet retained geometrical properties. It is much more plausible to think that all these properties can be attributed to an atom only with the same reservations, and such reservations may also later enable us to relate space and matter more closely. The two concepts, atom and empty space, would then no longer stand side by side yet be completely independent of one another; in this our atomic theory is even more consistent than that of the Greeks.

The second question will have to be discussed somewhat more fully. The statement that 'there are only atoms' meant to the Greeks, that all events, material and spiritual, must somehow be seen as movements of atoms. This would also apply to modern physics in-so-far as all processes are linked with changes in energy, and because of the atomic structure of energy, therefore linked with the movement of atoms. But the concepts 'soul' or 'life' certainly do not occur in atomic physics and they could not, even indirectly, be derived as complicated consequences of some natural law. Their existence certainly does not indicate the presence of any fundamental substance other than energy but it shows only the action of other kinds of forms which we cannot match with the mathematical forms of modern atomic physics. It follows that the mathematical structures of atomic physics are limited in their applicability to certain fields of experience and that, if we want to describe living or mental processes, we shall have to broaden these structures. It may be that we shall have to introduce yet other concepts which can be linked, without contradiction, with our existing systems of concepts. It may also become necessary to limit the range of previous concepts of atomic theory by attaching specific new conditions to them. In both cases we could regard such an extension as a broadened form of atomic theory and not as a theory describing only fundamentally different events.

If we accept such a wide definition of atomic theory we can immediately see how far removed we are from its completion. It would in fact amount to equating 'atomic theory' with a description of all reality, and this task will of course be infinite and will never be completed. We can only imagine a conclusion of atomic theory if we accept it in the limited sense I have sketched above. It would only have to deal with the special mathematical forms which serve for a description of the properties of elementary particles and the laws governing their transmutations at high energies. These mathematical forms may be very far-reaching but we cannot predict the magnitude of their range.

Even if we accept the second interpretation of the idea 'atomic theory', i.e. that 'there are only atoms and empty space' the materialism implied would in no way denote the anti-spiritual tendency which we commonly attach to this word. I hope that my previous explanations will have made that clear.

It may even be asked if we can still speak of materialism in this context. Let us seriously contemplate 'Just as tragedy and comedy can be written using the same letters so many varied events in this world can be realized by the same atoms as long as they take up different positions and describe different movements.'

It is important that we should understand the 'handwriting' of atoms for it is something which has not been thought out by man; it has far deeper meaning. Even when we shall have mastered and understood it, let us not forget that it is the content not the words which is important in a tragedy or comedy and that this also holds good for our world.

8

Science as a Means of International Understanding[1]

Dear Kommilitonen![2]

It has often been said that science should be a bridge between peoples and should help to better international understanding. It has also repeatedly been stressed, with full justification, that science is international and that it directs man's thoughts to matters which are understood by all peoples and in whose solution scientists of the most diverse languages, races or religions can participate equally. In speaking to you about this role of science at this particular time it is important that we should not make things too easy for ourselves. We must also discuss the opposite thesis, which is still fresh in our ears, that science is national and that the ideas of the various races are fundamentally different. It was held that science had to serve one's own people in the first instance and help to secure one's own political power: that science forms the basis of all technical developments, and hence of all progress, as well as of all military power. It was also held that the task of the pure sciences as well as of philosophy was to support our Weltanschauung and our beliefs. These in turn were regarded as the foundations of political power among our own people. I should like to discuss which of these two views is correct and what are the relative merits of the arguments that can be produced in their favour.

[1] Speech delivered before students of Göttingen University on July 13th 1946.
[2] This may be rendered as: Dear fellow graduates and under-graduates.

SCIENCE AND INTERNATIONAL UNDERSTANDING

I. To gain clarity on this question we shall have to discover, in the first instance, how science is carried on, how an individual is brought into contact with scientific problems and how these problems excite his interest. Since I know only my own science well, you will not misunderstand me if I first speak about atomic physics and if I recall my own experiences as a student.

When I left school in 1920 in order to attend at the University of Munich, the position of our youth as citizens was very similar to what it is to-day. Our defeat in the first world war had produced a deep mistrust of all the ideals which had been used during the war and which had lost us that war. They seemed hollow now and we wanted to find out for ourselves what was of value in this world and what was not: we did not want to rely on our parents or our teachers. Apart from many other values we re-discovered science in this process. After having studied a few popular books I began to take an interest in the branch of science concerned with atoms, and wanted to form an opinion of the peculiar statements which were being made about space and time in the theory of relativity. In this way I came to attend the lectures of my later teacher, Sommerfeld, who fanned this interest and from whom I learnt, in the course of the term, how a new and deeper understanding of atoms had developed as a result of the researches of Röntgen, Planck, Rutherford, and Bohr. I came to know that the Dane, Niels Bohr, and the Englishman, Lord Rutherford, imagined an atom to be a planetary system in miniature and that it was likely that all the chemical properties of the elements would, in future, be predictable with the help of Bohr's theory, by making use of the planetary orbits of the electrons. At that time, however, this had not been achieved. This last point naturally interested me most and every new work of Bohr was discussed at the Munich Seminar with vigour and passion. You can well imagine what it meant for me when Sommerfeld invited me, in the summer of 1921, to accompany him to Göttingen to hear a series of lectures given by Niels Bohr about his atomic theory. It was held in this very 'Collegienhaus'. This cycle of lectures in Göttingen, which in future was always to be referred to as the 'Bohr

SCIENCE AND INTERNATIONAL UNDERSTANDING

Festival', has in many ways determined my future attitude to science and especially to atomic physics.

First of all, we could sense in Bohr's lectures the power of the ideas of a man who had seriously grappled with these problems and who understood them better than anyone else in the whole world. Secondly, there were some points on which I had previously formed an opinion different from that expounded by Bohr. These questions were fought out during long walks to the Rohn and to the Hainberg.

These conversations left a deep impression on me. First I learnt that when trying to understand atomic structure it was obviously quite immaterial whether one was German, Danish or English. I also learnt something perhaps even more important, namely that in science a decision can always be reached as to what is right and what is wrong. It was not a question of belief, or Weltanschauung, or hypothesis; but a certain statement could either be simply right and another statement simply wrong. Neither origin nor race decides this question: it is decided by nature, or if you prefer, by God, in any case not by man.

Very much enriched by these experiences, I returned to Munich and continued, under Sommerfeld's direction, with my own experiments on atomic structures. When I had completed my Doctor's examination I went to Copenhagen, in the autumn of 1924, with the aid of a so-called Rockefeller Grant, in order to work with Bohr. There I came into a circle of young people of the most diverse nationalities—English, American, Swedish, Norwegian, Dutch and Japanese—all of whom wanted to work on the same problem, Bohr's atomic theory. They nearly always joined together like a big family for excursions, games, social gatherings and sports. In this circle of physicists I had the opportunity of really getting to know people from other nations and their ways of thought. The learning and speaking of other languages which this necessitated was the best way of becoming really familiar with other ways of life, foreign literatures and foreign art. I could see more and more clearly how little mattered the diversity of nations and races when there was common effort centred on a difficult scientific problem. The differences of thought which were so clearly shown in art seemed to me more of an enrichment of one's own possibilities than a disturbing factor.

With this background I arrived in Cambridge in the summer of 1925, and spoke about my work to a small circle of theoreticians in a College, in the study of the Russian physicist Kapitza. Among those present, there was an unusually gifted student hardly twenty-three years old who took my problems and constructed, within a few months, a comprehensive theory of the atomic shell. His name was Dirac and he was a man of outstanding mathematical ability. His methods of thought were vastly different from mine, his mathematical methods more elegant and more unusual than those to which we were used at Göttingen. However, in the end, he arrived at the same results as Born, Jordan and I, at least on all points of importance. This confirmation and the fact that the results were so beautifully complementary served as further proof of the 'objectivity' of science and its independence of language, race or belief.

As well as Copenhagen and Cambridge, Göttingen remained a centre for this international family of atomic physicists. The work was directed by Franck, Born and Pohl and many of the scientists about whom you read in the newspapers in connection with the atom bomb, such as Oppenheimer and Blackett, as well as Fermi who studied in Göttingen at that time.

I have quoted these personal reminiscences only in order to give an example of the internationalism of the community of science. It has, of course, been the same for centuries in many other sciences and this family of atomic physicists was in no way out of the ordinary. I could quote many international groups of 'savants' from the history of science who were linked through the frontiers of nations by common work.

Perhaps I might mention one other group of scientists who, in the seventeenth century, founded mathematical science in Europe. It is especially appropriate to do so because the memory of Leibnitz is being celebrated this year as well as the foundation of the Scientific Academies. I should like to quote a few sentences of Dilthey's description of that epoch.

SCIENCE AND INTERNATIONAL UNDERSTANDING

'A bond, unhampered by any limitations of language or nationality, linked the few individuals who devoted their lives to this new science. They formed a new aristocracy and were conscious of it, just as before in the days of the Renaissance, humanists and artists had felt themselves to be such an aristocracy. The Latin and, later on, the French language rendered the easiest mutual understanding possible and they became the instrument of a scientific world literature. Already around the middle of the seventeenth century, Paris had become the centre of collaboration between philosophers and scientists. There Gassendi, Marsenne and Hobbes exchanged ideas and even the proud recluse Descartes joined their circle for a time. His presence made an unforgettable impression on Hobbes and later Leibnitz; for it was there that both became devoted to the ideas of mathematical science. Later, London became another centre. ...'[1]

We can see then that science has been carried on in this way throughout history and that the 'Republic of Sages' has always played an important part in the life of Europe. It has always been considered self-evident that adherence to such an international circle would not prevent the individual scientist from devotedly serving his own people and feeling himself one of them. On the contrary, such a broadening of one's horizon frequently enhances esteem for the best aspects of the life of one's own country. One learns to love it and feels indebted to it.

II. Having said all this I must now also deal with the question of why all this scientific collaboration, all these real human relationships, seemingly do so little in preventing animosity and war.

First of all it must be stressed that science represents only a small part of public life and that only very few people in each country are really connected with science. Politics, however, are shaped by stronger forces. They have to take into account the actions of large masses of people, their economic position and the struggle for power of a few privileged groups favoured by

[1] Dilthey: Gesammelte Werke Bd. III, S.15, 16.

tradition. These forces have, so far, always overpowered the small number of people who were ready to discuss disputed questions in a scientific way—that is, objectively, dispassionately and in the spirit of mutual understanding. The political influence of science has always been very small, and this is understandable enough. It does, however, frequently place the scientist in a position which is in some ways more difficult than that of any other group of men. For science has, in its practical applications, a very great influence on the life of the people. Prosperity and political power depend on the state of science and the scientist cannot ignore these practical consequences even if his own interest in science is of a less practical nature. Thus, the action of an individual scientist often carries far more weight than he would wish and he frequently has to decide, according to his own conscience, whether a cause is good or bad. When the differences between nations can no longer be reconciled he is therefore often faced with the painful decision either of cutting himself off from his own people or from those friends who are linked with him by their common work. The position in the various sciences is here somewhat different. The medical practitioner, who helps people irrespective of their nationality, can more easily reconcile his actions to the demands of the state and of his own conscience than the physicist, whose discoveries may lead to the manufacture of weapons of destruction. But, by and large, there always remains this tension; there are on the one hand the demands of the state, which wants to enlist science particularly for the benefit of its own people and hence the strengthening of its own political power. On the other hand there is the duty owed by the scientist to his work which links him to people of other nations.

The relations between the scientist and the state have changed in a characteristic way during the past decades. During the first world war the scientists were so closely tied to their states that Academies frequently expelled scientists of other countries or signed resolutions in favour of their own cause and against the cause of the other nation. This hardly happened at all during the second world war. The link between the scientists was frequently much stronger, even to the extent, in many countries, of difficulties arising between them and their own governments. Scientists claimed the right to judge the policies of their governments independently and without ideological bias. The State, on its side, viewed the international relations of scientists with deep mistrust so that eventually scientists were sometimes even treated like prisoners in their own country and their internattional relations considered almost immoral. Conversely it has now become almost a matter of course that scientists will help their colleagues wherever possible, even though they belong to the enemy country. This development may lead to a fortunate strengthening of international, as against national, relations, but care will have to be taken that it does not become the origin of a dangerous wave of mistrust and enmity of large masses of people against the profession of science itself.

There have been such difficulties in previous centuries when men of science stood up for the principle of tolerance and independence from dogma against the current political power. We need only think of a Galileo or a Giordano Bruno. That these difficulties have assumed even greater importance to-day may be because the practical effects of science can directly decide the fate of millions of people.

This brings me to a frightening aspect of our present-day existence which has to be clearly recognized so that the correct action can be taken. I am not only thinking of the new sources of energy which physics has mastered during the last year and which could lead to unimaginable destruction. New possibilities of interfering with nature are threatening us in many other fields, though it is true that chemical means of destroying life have hardly been used in this last war. In biology, too, we have gained such insight into the processes of heredity and into the structure and chemistry of large albumen molecules that it has become a practical possibility to produce infectious diseases artificially, and perhaps worse, even the biological development of man may be influenced in the direction of some predetermined selective breeding. Finally, the mental and spiritual state of people could be influenced and, if this were carried out from a scientific point of view, it could lead to terrible mental deformations of great masses of people. One has the impression that science approaches on a broad front a region in which life and death of humanity at large can become dependent on the actions of a few, very small groups of people. Up to now these things have been discussed in a journalistic and sensational way in the newspapers and most people have not realized the terrible danger which threatens them as a result of further inevitable scientific developments. It is certainly the task of science to rouse humanity to these dangers and to show them how important it is that all mankind, independent of national and ideological views, should unite to meet the peril. Of course, this is more easily said than done, but it is certainly a task which we can no longer escape.

For the individual scientist there remains, however, the necessity of deciding according to his own conscience and free from all ties, whether a cause is good or even which of two causes is less bad. We cannot escape the fact that large masses of people, and with them those who hold the power of government, often act senselessly and with blind prejudice. By giving them the scientific knowledge the scientist can easily be manouvered into a position which Schiller describes in these verses:

'Woe to those who bestow the light of heaven on him who is for ever blind, it sheds no light for him, it can but char and blacken lands and cities.'

Can science really contribute to understanding between the peoples when it is faced with such a situation? It has the power to release great forces, greater than have ever before been in the control of man, but these forces will lead into chaos unless they are sensibly used.

III. This leads me to the real inherent task of science. The development which I have just described and which has apparently turned against himself those forces which man controls and which can lead to the most terrible destruction, this development must certainly be closely connected with some spiritual processes of our time, and it is necessary to speak briefly about these.

Let us look back a few centuries. At the end of the Middle Ages man discovered, apart from the Christian reality centred round the divine revelation, yet another reality of material experience. That was 'objective' reality which we experience through our senses or by experiment. But in this advance into a new field certain methods of thought remained unchanged. Nature consisted of things in space which changed in time according to cause and effect. Outside of this there was the world of spirit, that is, the reality of one's own mind which reflected the external world like a more or less perfect mirror. Much as the reality determined by the sciences differed from the Christian reality, it nevertheless represented also a divine world order with man's action based on a firm foundation, and in which there could be little doubt about the purpose of life. The world was infinite in space and time, it had in a way replaced God or had at least become, by its infinity, a symbol of the divine.

But this view of nature has also become undermined during our century. Fundamental attitudes of thought lost their absolute importance as concrete action moved more and more into the centre of our world. Even time and space became a subject of experience and lost their symbolic content. In science we realize more and more that our understanding of nature cannot begin with some definite cognition, that it cannot be built on such a rock-like foundation, but that all cognition is, so to speak, suspended over an unfathomable depth.

This development of science corresponds probably to the increasingly relative assessment of all values in the life of man, an assessment which has been noticeable for some decades and which can easily end up in a generally sceptical attitude capped by the desperate question 'for what purpose'. Thus develops the attitude of unbelief which we call 'nihilism'. From this point of view life appears to be purposeless or, at best, an adventure which we have to endure while having had no say in it. We find this attitude in many parts of the world to-day and its most unpleasant form is illusionary nihilism, as v. Weizsäcker recently called it. It is a nihilism disguised by illusion and self-deception.

The characteristic trait of every nihilist attitude is the lack of a solid belief which can give direction and strength to all the reactions of an individual. Nihilism shows itself in the life of an individual by his lack of an unerring instinct for right and wrong, for what is an illusion and what is a reality. In the life of nations it leads to a change of direction in which the immense forces, which have been gathered for the achievement of a certain aim, have the very opposite result and this can cause great destruction. People are often so blinded by hatred that they cynically watch this change and dispose of it with a shrug of the shoulder.

I said a little earlier that this development in the outlook of men may have some relation to the development of scientific thought. We must therefore ask whether science too has lost its solid beliefs. I am very anxious to make it quite clear that there can be no question of this. The very opposite is true. The present situation of science is probably the strongest argument we possess for a more optimistic attitude to the great problems of the world.

For in *those* branches of science in which we have found that our knowledge is 'suspended in mid-air' in *just those* branches have we achieved a crystal clear understanding of the relevant phenomena. This knowledge is so transparent and carries such force of conviction that scientists of the most diverse peoples and races have accepted it as the undoubted basis of all further thought and cognition. Of course, we also make mistakes in science and it may take some time before these are found and corrected. But we can rest assured that there will be a final decision as to what is right and what is wrong. This decision will not depend on the belief, race or origin of the scientists, but it will be taken by a higher power and will then apply to all men for all time. While we cannot avoid in political life a constant change of values, a struggle of one set of illusions and misleading ideas against another set of illusions and equally misleading ideas, there will always be a 'right or wrong' in science.

SCIENCE AND INTERNATIONAL UNDERSTANDING

There is a higher power, not influenced by our wishes, which finally decides and judges. The core of science is formed, to my mind, by the pure sciences, which are not concerned with practical applications. They are the branches in which pure thought attempts to discover the hidden harmonies of nature. Mankind to-day may find this innermost circle in which science and art can hardly be separated, in which the personification of pure truth is no longer disguised by human ideologies and desires.

You may, of course, object that the great mass of people has no access to this truth and that it can therefore exert little influence on the attitude of people. But at no time did the great mass of people have direct access to the centre and it may be that people to-day will be satisfied to know that though the gate is not open to everyone there *can* be no deceit beyond the gate. We have no power there—the decisions are taken by a higher power. People have used different words at different times for this 'centre'. They called it 'spirit' or 'God', or they spoke in similes, or in terms of sound or picture. There are many ways to this centre, even to-day, and science is only one of them. Perhaps we have no longer a generally recognized language in which we can make ourselves intelligible. That may be the reason why so many people cannot see it, but it is there to-day as it has always been, and any world order must be based on it. Such a world order must be guided by men who have not lost sight of it.

Science can contribute to the understanding between peoples. It can do so not because it can render succour to the sick, nor because of the terror which some political power may wield with its aid, but only by turning our attention to that 'centre' which can establish order in the world at large, perhaps simply to the fact that the world is beautiful. It may appear presumptuous to attribute such importance to science but may I remind you that though we have cause to envy previous epochs in many aspects of life, our age is second to none in scientific achievement, in the pure cognition of nature.

Whatever may happen, interest in knowledge itself will remain a potent force in mankind for the next few decades. Even though this interest may for some time be overshadowed by the practical consequences of science and by the struggle for power it must eventually triumph and link together people of all nations and races. In all parts of the world people will be happy when they have gained new knowledge and they will be grateful to the man who first discovered it.

Dear Kommilitonen, you are gathered here to contribute in your circle to an understanding between the peoples. There can be no better way of doing this than by getting to know, with the freedom and spontaneity of youth, people of other nations, their ways of thought and their feelings. Take from your scientific work a serious and incorruptible method of thought, help to spread it, because no understanding is possible without it. Revere those things beyond science which really matter and about which it is so difficult to speak.

Index

Abdera, 97
Abgeschlossenheit, 24
'Absolute' time, 20, 40, 47
'Accidental' mechanical effects, 74–5
'Analytical' and 'immediate' concepts, 30
Anaxagoras, 28–9, 96–7
Anaximander, 96
Anderson, on matter and radiation, 51, 102
Anschauungsformen, 20
Archimedes, 42
Aristotle on 'empty space', 30–1, 34
Astronomy and animate processes, 81
Astro-physics, 88
Atom-hypothesis, and chemistry, 81–4
Atomic: apparatus, 19; nuclei, 100–1, (*and see* Electrons, Protons, etc.); phenomena and classical physics, 23 (*and see* Classical physics, Greek thought); physics, early, 28 seqq.; physicists, future aims of, 95 seqq.; structures, Bohr's theory of, 14 (*and see* Bohr; theory, ultimate goal of, 105
Atomization of science, 79–80
Atoms: and non-material fields, 106–8; 'perceptible' qualities of, 55

Basic: chemicals, reduction of, 101; 'elements' conception, 98–9
Being and becoming, polarity of, 96
Being and non-being, 55
Biology and atomic considerations, 82, 91
Blackett, Professor, 112
Bohr, Niels, 14, 15, 20, 25, 38, 39, 41, 84, 85, 89, 99, 110, 111
'Bohr Festival', 110–1
Bolzmann, 16
Born, 14, 112
Bothe, 17
'Boundary' in colour production, 61
Brahe, Tycho, 19
Broglie, 14
Bruno, Giordiano, 115

Cambridge, 112
Cell-division, 91, 92
Chemists' methodology, 80, 81, 82, 83, 84–8
Classical physicists: and prediction, 50, 51; and 'mental field', 92
Classical physics: atomic phenomena in, 23–4; basis and

INDEX

validity of, 41, 42, 43, *and see* Newton, etc.; birth of, 21–2; imprecise concepts of, 42–3; fundamentals of, 11, 12, 13, 14, 19, 20, 21, 25, 26; laws and concepts of, compared, 45; scope of, 23–4; truth content of, III passim

'Cogito, ergo sum', 25

Colour; Goethe's and Newton's approach to, 36, 37, 38, 39, 60–76; and number, 63. *See also* White light

Columbus, C., 17, 18, 23, 25

Complete comprehension, ideal of, 26

Compton, 17

Contemplation of Nature (Goethe), 61

Copenhagen, 111–2

Copernicus, 13, 14, 17, 35, 78

Corpuscular concept (light and matter), 14

Cosmic harmony (Kepler's), 78

Crystals, mathematics applied to, 57, 89

Darwin, C., 81

Democritus, 27, 29, 30, 31, 37, 38, 54, 55, 97, 106

Descartes, 22, 25, 113

Destructive possibilities of atomic physics, 115–7

Differential equations, *see* Newton

Dilthey, 112

Dingler, 45

Dirac, 15, 51, 52, 112

'Disturbance', 15–6, 73

'Dividing line', 15, 16; and statistics, 49, 50

Dynamic laws, mathematics and, 57–8

Einheitlichkeit, 53–4

Einstein, relativity theory of, 12, 13, 14, 20, 42, 44, 45, 46, 47 48, 102

'Elan vital', 80

Electric field concept, 69

Electric-magnetic phenomena, 6

Electrolysis, 38

Electrons, 54, 99, 100, 101, 102, 103; deflection of, 49; limitation of concepts through discovery of, 51; positions of, 51; protons, neutrons and, 103

Elementary particles, 102, 103

Empedocles, 28, 97

'Empty space' concept, 30, 31, 54, 55, 97, 106

'End of world in space' concept, 17, 23, 24

Energy as 'fundamental substance', 103

Erkenntnistheoretische Analyse, 34

Euclid, 13, 20, 45

'Exact science', 80, 81, 88; changing foundations of, 11–26

Experiments (atomic), situation behind, 49

'False' protons, 46, 47, 52

Faraday, M., 52, 99

Fichte, 28

Fields, measuring problems in, 51

Finite range of science, 73–4

Five atoms of Plato, 98

Four stages of perception (Plato), 32–4

Franck, 14

'Fundamental substance', 53, 54 (*see also* Greek thought, Thales).

122

INDEX

Galileo, 13, 34-5, 66, 77, 78, 115
Gassendi, 37, 113
Geiger, 17
Geometrical applications, 22, 29, 30, 38, 39, 57, 72, 97, 98
Gerlach, 14
Gesetzmässigkeiten, 48, 50
Goethe, 37, 60-76; and Newton, on colour, 60 seqq.
Göttingen circle, 15, 110-1
Greek thought (ancient theories of physics), 28-34, 52 seqq., 96 seqq., 106

Hahn, 101
'Handwriting' of atoms, 108
Harmonies: of mathematics, 56, 57; of spheres, 67; Pythagorean, 56
Harmonization, 51, 52
Heat, as substance, 37
Heat exchange, 89
Heat-mechanics link, 83
Helmholtz, on Goethe, 70
Heraclitus, 96
Hertz, H., 14
Hobbes, 113
'Holes' theory (Dirac's), 52
Huygens, 36

Illusionary nihilism, 118
Imagination, 30
Inconsistency inherent in atomic theory, 54-5
Indestructibility, 53, 54, 97
Indivisible particles, 54, 55
Inevitability of progress, 71-3
International understanding, and science, 111-20
Intuition and science, 64

Jordan, 112

Jowett, B., 32 n.

Kaleidoscope, 56, 65
Kant, E., 20, 21, 22
Kapitza, 112
Kepler, 19, 36, 57, 66, 67, 77, 78, 94
Kramers, 14

Language, limitations of, 43, 44, 67
Laplace, 80
Laws of causality, 20
Leibniz, 112
Lenard, 14
Leucippus, 27, 29, 54, 97
Light, 14, 36, 102; as electromagnetic phenomenon, 69; plus dark (Goethe), 62
'Living' aspects of nature, 74
Loyalty, physicists' problems of, 114

Macroscopic matter, 83, 84, 85, 86
Magellan, 17
Magnetism, 69
Marsenne, 113
Materie im Grossen, *see* Macroscopic matter
Materialism, 30, 31, 56, 97, 106, 107, 108
Mathematics, II passim; as heuristic principle, 58-9; future role of, in nuclear physics, 104; crystals and, 57; Goethe and, 64; Kaleidoscope and, 57; music and, 56-7; optics and, 66; purposive force of, 53, 56
Matter, 103; and radiation, 51; post-Greek study of, 37 seqq.;

INDEX

sense-properties of, 70. *See* Space
Maxwell's equations, 18, 31, 36, 37, 52, 83
Measuring apparatus, atomic structure of, 52
Mechanistic school, 89–90
Mental processes and science, 82, 92–3
Mesons, 102, 103
Metamorphosis of Plants (Goethe), 61
Michelson-Morley experiments, 12
Mixing phenomena, early notions of, 54
Modern physics: basis of, 42; 'false' problems and, 46, 47; fuller expansion of, 11 seqq.; Goethe and Newton and, 60–76; natural philosophy of the ancients and, 53–9; necessary revision of, 51–2; principle in, 41–52
Modern science, dangers of, 71 seqq.
Molecules, 84, 99
Monochromatic light ray, 68
Music, 56–9
Mutations and statistical laws, 91

Natural philosophy; modern physics and, 53–9; religion and, 21
Naturer-klärung and -beschreibung, 34
Nature: changing concepts of, 117; physical interpretation of, 27 seqq.
Neutrons, 54
Newton, 31, 35, 36, 39, 41, 42, 57, 60 seqq., 80, 87; on changing attitude to science, 79
Nietzsche, 96
Nihilism, 117-18
Nineteenth-century rationalism, 21, 22, 23, 24–6

Objectivity, 64, 68, 87, 112, 117
Observer and observed object, 12, 15, 16, 19, 38
Oken, Lorenz, 36
Oppenheimer, 112
Optics, 36–7, 63, 66, 83

Parmenides, 28–30; 'imagination' and, 30; *Seins* idea and, 28
Past, present, future idea, 12 seqq.
Pauli, 14
Philosophy, early approach to, 21–3; mathematical balance of, 40
Philosophy, perception and, 20, 21
Physical interpretation of nature, 27–46
Physical optics, Newton's theory of, 62 seqq.
Physics, 11, 18–19, 24; *see* Classical, Modern physics
Planck, M., 11, 14, 84–5, 100, 102, 110
Planetary motion; Brahe's and Kepler's work on, 19; old and new concepts of, 35–6
Plato, 30–3, 34, 35; *Dialogues* of, 33n., 35n., and *Timaeus*, 30, 35, 57, 98
Pohl, 112
Polar relations and colour arrangement, 65

INDEX

Politics and physicists' work, *see* Science
Positive electron (Anderson's), 102
Position, 52
Principle in modern physics, 41–52
Protons, 54, 100, 101, 102
Prout's hydrogen hypothesis, 99
Ptolemy, 13
Pure science, role of, 20, 116 seqq.
Pythagoras, 31, 33, 56, 59

Quantum laws and theory, 11, 14, 15, 16, 17, 18, 20, 24, 25, 41, 44, 45, 46, 47, 48, 84, 85–6, 89, 93, 96; and wave function, 49, 50; laws of causality in, 20 seqq.; relativity harmonizing with, 51; statistical statement of, 50, 51

Radiation, 51, 52, 62, 63, 102
Radio-activity, 50
Rationalism, classical physics and, 22, 23, 24
Refraction and colour effects, 61
'Related aspects' (Goethe's), 74
Relativity, 12, 13, 14, 20, 42, 44, 45, 46, 47, 48, 102
Religion, *see* Natural Philosophy
Renaissance, 21
Research, broadening of spheres of, 79
'Romantic' natural philosophy, 81
Röntgen, 110
Rutherford, E., 14, 85, 99, 110

Schrödinger, 15, 48, 101
Sciences: complementary character of, 20; international nature of, 77–94, 111–20; national considerations and, 109 seqq.; politics and, 113, 114; real tasks of those engaged in, 116 seqq.; right and wrong in, 118, 119
Seins, *see* Parmenides
Selbstbeschränkung, 28
Sense properties of matter, 70
Simultaneity, 12, 46
Sommerfeld, 14, 51, 110, 111
'Soul' and atomic physics, 107
Space and matter, 30 seqq.
Space and time, 11, 12, 13, 14, 15, 30
Specialization, 79–80
Spinoza, 22
Stammwurzeln, 97
Stark, 14
Stern, 14
'Stossen und Schlagen', 36, 81
Substance, fundamental, 28 seqq., 53, 54
'Symbolic' atom concept, 56

Tasks of science, 18, 116 seqq.
Technological progress, inevitability of, 71–3
Thales of Miletus, 28, 29, 33, 96, 97; 'water' theory of, 53, 96
Thermodynamics, kinematic interpretation of, 50
Transformation of particles, 102, 103

Unity of Science, 77–94
Urphänomen (Goethe's), 62, 63

Valency, theory of, 88
Vitalism, 80; and mechanism, 88–90

Wärmehaushalt, *see* Heat exchange
'Water' origin, *see* Thales of Miletus
Wave function, 15, 49
Weizsäcker, 118
Weltanschauung, 25

White light, 62, 63
Wilson cloud chamber, 42, 43, 46
Wöhler, 81

X-radiation, 102

Zerstrahlung, 52